SpringerBriefs in E

More information about this series at http://www.springer.com/series/10184

David Koepsell

Scientific Integrity and Research Ethics

An Approach from the Ethos of Science

☺ Springer

David Koepsell
Research and Strategic Initiatives at
 Comisión Nacional de Bioética
 (CONBIOETICA) Mexico
Universidad Autonoma Metropolitan –
 Xochimilco
Mexico City, Mexico

The Work was first published in 2015 by Comisión Nacional de Bioética (CONBIOETICA) with the following title: Ética de la Investigación, Integridad Cientíca.

ISSN 2211-8101 ISSN 2211-811X (electronic)
SpringerBriefs in Ethics
ISBN 978-3-319-51276-1 ISBN 978-3-319-51277-8 (eBook)
DOI 10.1007/978-3-319-51277-8

Library of Congress Control Number: 2016962457

© Springer International Publishing AG 2017
This work is subject to copyright. All rights are reserved by the Publisher, whether the whole or part of the material is concerned, specifically the rights of translation, reprinting, reuse of illustrations, recitation, broadcasting, reproduction on microfilms or in any other physical way, and transmission or information storage and retrieval, electronic adaptation, computer software, or by similar or dissimilar methodology now known or hereafter developed.
The use of general descriptive names, registered names, trademarks, service marks, etc. in this publication does not imply, even in the absence of a specific statement, that such names are exempt from the relevant protective laws and regulations and therefore free for general use.
The publisher, the authors and the editors are safe to assume that the advice and information in this book are believed to be true and accurate at the date of publication. Neither the publisher nor the authors or the editors give a warranty, express or implied, with respect to the material contained herein or for any errors or omissions that may have been made. The publisher remains neutral with regard to jurisdictional claims in published maps and institutional affiliations.

This Springer imprint is published by Springer Nature
The registered company is Springer International Publishing AG
The registered company address is: Gewerbestrasse 11, 6330 Cham, Switzerland

*Dedicated to Richard Hull and David Triggle,
Mentors and friends in research ethics*

Contents

1 Introduction to Scientific Integrity and Research Ethics 1
    1.1 What Is "Science" and Its Ethos? 1
    1.2 Early Lapses in Science and Ethics 2
    1.3 Nuremberg and Its Progeny 3
    1.4 The Origins of the Nuremberg Principles 5
        1.4.1 Virtue Ethics 5
        1.4.2 Deontology 6
        1.4.3 Consequentialism/Utilitarianism 7
    1.5 Some Problems with Ethical Theories 8
    1.6 The Modern Bioethics Regime 9
    References 12

2 Research (Mis)Conduct 13
    2.1 Sources and Citations 13
    2.2 Data Collection, Manipulation, and Curation 16
    2.3 Correlation Without Causation 18
    2.4 Publication Bias as Part of the Problem 20
    2.5 Fraud 21
    2.6 Conclusions 22
    References 22

3 Issues of Authorship 25
    3.1 Publish or Perish 25
    3.2 Duties to the Truth 26
    3.3 Duties of Authors to Each Other 27
    3.4 Are You an Author? 28
    3.5 The Korean Stem Cell Fraud 30
    3.6 What Counts as Your Work? 32
    3.7 Salami Science and Self-Plagiarism 34
    3.8 Conclusions 34
    References 35

4 Issues in Intellectual Property and Science ... 37
4.1 What Is Intellectual Property (IP)? ... 37
4.2 A Short and Sweet History of IP ... 38
4.3 Who Owns What? ... 41
4.4 Not Treading on IP During Research ... 42
4.5 Science and Competition ... 43
4.6 Nature vs. Artifacts: What Ought to Be Monopolized Consistent with the Ethos of Science? ... 44
4.7 Open Science as an Alternative ... 45
4.8 Conclusions ... 46
References ... 47

5 Conflicts of Interest ... 49
5.1 What Is a Conflict of Interest? ... 49
5.2 The Interests of Science ... 50
5.3 Interests of Scientists ... 52
5.4 Other Institutional Interests ... 53
5.5 Equipoise: A Duty of Scientists and Their Institutions ... 54
5.6 The Problem of Private Ethics Committees and Contract Research Organizations ... 55
5.7 The Gelsinger Case ... 56
5.8 What Can Be Done? ... 58
References ... 59

6 Autonomy, Dignity, Beneficence, and Justice ... 61
6.1 The Emergence of Medical Ethics ... 61
6.2 Autonomy ... 63
6.3 Dignity ... 65
6.4 Beneficence/Non-maleficence ... 66
6.5 Justice ... 67
6.6 The Four Principles and "Care" Ethics ... 69
References ... 71

7 Ethics Committees: Procedures and Duties ... 73
7.1 Legal and Regulatory Framework ... 73
7.2 Some Best Practices: Ethics Committees in Biomedicine ... 74
7.3 Minimizing Risks: Stakeholders and Duties ... 76
7.4 Clinical Trials: Methods and Duties ... 78
7.5 Randomizing and Blinding ... 80
7.6 Some Ethical Issues in Clinical Trials: Risk vs. Benefit ... 81
7.7 Informed Consent ... 82
7.8 Vulnerability and Justice ... 83
References ... 83

8	**Duties of Science to Society (and Vice Versa)**	85
	8.1 Science and Society	85
	8.2 Universalism	86
	8.3 Communalism	88
	8.4 Disinterestedness	89
	8.5 Organized Skepticism	91
	8.6 Some Conclusions and Some Remaining Questions	92
	References	94

Appendix: Codes and Principles 97
The Nuremberg Code 97

Declaration of Helsinki 99
Recommendations Guiding Doctors in Clinical Research 99
    INTRODUCTION 99

The Belmont Report 103

Table of Contents 105
Ethical Principles & Guidelines for Research Involving
Human Subjects 105
Part A: Boundaries Between Practice & Research 106
  A. Boundaries Between Practice and Research 106
Part B: Basic Ethical Principles 107
  B. Basic Ethical Principles 107
Part C: Applications 110
  C. Applications 110

Introduction

Science is one of our most important modern institutions. Its steady progress, by which we have gained a better understanding, greater ability of prediction, and ever-increasing control over our world, has improved our lives in countless ways. As a human institution, it is fallible. As an institution embedded and dependent upon numerous other institutions, it is prone to pressures from those who are not necessarily "part" of scientific endeavors per se. Instances of what we might call scientific misconduct, breaches of research ethics, and failures of scientific integrity are not new. This book refers to a number of cases from ancient times to modern which threatened science, undermined its norms, and which we might call unethical. Many scholars have examined and tried to understand the nature of scientific misconduct, why it occurs, and what makes it wrong. Every prominent instance of such misconduct jeopardizes both the progress of science and the public's perception of scientists and their institutions. Given its importance and its unqualified success in extending and improving the quality of our lives, we ought to be mindful of how we, as researchers, present ourselves and our work to the public and avoid where possible both the appearance and occurrence of wrongdoing.

When the public was less aware and generally poorly informed about science and the academy, failures of scientific integrity may have been less harmful to the health of scientific institutions and to their general political support. But in a world of increasingly available knowledge, and in which a greater emphasis is given to democratic processes, the dangers of lapses of scientific integrity and research ethics are increased. If the public loses confidence in our research institutions, then the already threatened fiscal and emotional support for its members and programs risks being further jeopardized. We should all be concerned with preserving the integrity of the institutions of science and recognize the role of each researcher in maintaining that integrity. How can we do so? Even as we research the theory behind research ethics, the potential moral bases for good conduct in science, and the sociology of scientific institutions, there are numerous cases and examples that, when seen in their historical context, provide us with illustrations of how misconduct impacts science and individuals. We can start by looking at those cases and appealing to what we know about the way science works when it isn't pathological.

In 2004, I began teaching a research ethics course for graduate students at the University at Buffalo with David Triggle. He and Richard Hull, a mentor and colleague from UB, had developed and were teaching it, mostly for students in the pharmacy faculty, but also occasionally students from other disciplines. Dr. Triggle introduced me to the use of Merton's ethos of science as a basis for instruction of scientific integrity. He used Merton in one of his lectures, and I came over time to adopt it for a range of cases and behaviors which we associate with both good scientific practice and ethical conduct. I took over full responsibility for the graduate research ethics course at UB in 2006 and attempted to try to expand its usefulness and start to include students from other faculties. At the time, only pharmacy graduate students were required to take the course, though we were always reviewed very highly and students and professors considered what we were teaching to be valuable.

For two years I attempted, with the help and support of colleagues at UB who were enthusiastic about expanding the program, to bring in faculties and students across the university. The general goal was to help foster an environment of research ethics and increased scientific integrity for all disciplines. Joe Gardella and Bruce Pitman especially were supportive and helped to try to rally the support of the central administration and other faculties. Like many such efforts, where already overtaxed graduate students might be required to attend more courses rather than attend to their duties in the lab, there was some resistance. Before I could secure any commitments, I was offered a tenure-track job at Delft University of Technology in the Netherlands and jumped at the opportunity.

Shortly after arriving in Holland, the "Stapel affair" (Diedrick Stapel was involved in an enormous case of academic fraud that has been well-publicized) brought a great deal of attention to issues of scientific integrity in Holland. Delft, which had recently developed a code of ethics and academic integrity (which I was asked to help draft), became very concerned about increasing its commitment to instruction in research ethics. Their response was on two fronts, including a mandatory three-hour workshop for all faculty (in small groups at a time) over the course of the year and a six-hour workshop for all incoming Ph.D. students within the first year of study. I helped design and teach both programs, and the rector magnificus himself, Prof. Karel Luyben, took part in the faculty workshop sessions. Both programs were very well received, and we shall see over time the degree to which such training may or may not increase actual adherence to norms of scientific integrity. I hope and believe they will.

In 2013, my wife secured a position back in her home country of Mexico as a researcher at INMEGEN, Mexico's national center for genomic research, and so my family and I went to Mexico. There, I became director of Research and Strategic Initiatives at CONBIOETICA, Mexico's National Commission of Bioethics. As part of Mexico's increasing concern and commitment to an environment of scientific integrity, CONBIOETICA had received a grant from the Mexican national science foundation (CONACYT) to publish a book on the subject. I was asked to draft the book and eagerly leapt at the opportunity. I decided to model the book after the lectures I had delivered over the past decade and focus on the approach I had developed in no small part thanks to my work with David Triggle and the teaching I had done over ten years.

Most of my students in research ethics courses at UB, Delft, and now in Mexico have come from the sciences. Only a handful have been students in philosophy or other humanities, with the great majority being from the hard and social sciences. Early on in teaching such students, I realized that while some brief and shallow background in ethical theory was helpful for developing context and vocabulary, the general appeal to the Mertonian norms of science was a more effective means of reaching the scientifically minded and philosophically skeptical. I gradually began to emphasize those norms, even while continuing to teach some basics of ethical theory, as a means of appealing to those who may be more inclined to support the institutions of science for their own sake, rather than appeal to deeply considered, though less empirically supported, theories devised by philosophers.

Philosophers have debated ethical theory for millennia, and there is still no general agreement among philosophers of ethics as to how to determine the good. Unlike scientific progress, which gets buttressed over time by accumulating evidence, testable, and also falsifiable by new evidence, philosophical theories about the good cannot be measured or tested with similar objectivity. I think that research into ethical theory is an essential philosophical topic, and university philosophy departments wisely continue to hire and fund researchers in those areas, but it is not particularly worthwhile to dwell too deeply in those issues for the sake of preparing researchers to conduct research ethically or to abide by norms of academic and scientific integrity. Nor is it really necessary, in my opinion.

We should be able to assume that those who become academics or researchers have some concern with the science itself, with research qua research, with the general pursuit of truth and a better understanding of the universe and our place in it afforded by adherence to the general principles of empiricism, the scientific method, and the institutions that have evolved around science. Of course, we are all embedded in a number of institutions simultaneously. Our communities, polities, families, faiths, etc., all demand our attention and sometimes conflict with one another. It is from that basic presumption that every researcher is at some level concerned with science itself that I suggest we can all as researchers uniformly proceed to develop norms of behavior without recourse of the vagaries of philosophical ethics.

For science to work, scientists must necessarily comport themselves with certain behaviors, regardless of their motives. Science is the pursuit by diverse researchers, often in geographically distant locales, crossing cultural barriers and belief systems, of universal truths. Science only works, meaning it progresses us closer to an accurate understanding of phenomena, if we all accept its universalism. It also only works correctly if we understand that it is a communal pursuit and not a solitary one. We must also all accept that the current status of all science is contingent, so we remain skeptical and willing to set aside cherished theories in light of new evidence. Similarly, we must not be emotionally invested in our studies, but rather be always willing and ready to be shown to be wrong. This equipoise or disinterestedness and other norms work together to make science a highly successful institution as a whole, even while failures and dead ends abound. It is from this starting point, from an acceptance of the necessity of the norms of science itself, that we can begin to examine what it means to conduct oneself according to scientific integrity and form a working theory of what ethical research looks like.

This short book is intended to be a useful resource in studying and discussing issues in scientific integrity and research ethics. A number of fine resources exist in this field, including Francis Macrina's textbook which I have used in teaching for many years. This text, I hope, supplements the existing corpus by providing a compact introduction to the main themes of scientific integrity and research ethics suitable for any audience, especially scientists and other researchers rather than students of ethics or philosophy of science.

My approach to the subject has changed over time, and this text embodies what I think is a way to discuss scientific integrity using cases but guided by a coherent philosophy which does not itself require ascribing to a particular ethical system. While I will introduce philosophical concepts of ethics, grasping them or the debates surrounding them over time is unnecessary to developing a method of discerning, by and large, the correct responses to conundrums in research ethics as they arise. By correct, I don't mean morally correct as we will abide by the notion that moral truths are currently unknown and maybe unknowable, but rather when science proceeds according the norms or ethos of science itself, and scientists comport themselves according to those norms, fewer harms will presumably occur, and scientific progress will be accentuated. I assume science is, as a whole, being pursued ethically and that most researchers are careful and concerned about their work and how it impacts humanity. I also assume that errors happen and are generally innocent. But it is from these assumptions that I proceed with the hope that we can all afford to educate ourselves about the state of the art as well as long history of scientific and academic study, errors, and harms and guide our behavior by some set of shared principles in the hope we can avoid similar harms and errors.

I hope this book is useful, and at the end of each chapter I provide some questions that may be used either in pedagogy or for self-reflection to help to consider some of the issues and cases presented and to sharpen thinking about issues as they may arise. I am grateful for the opportunity to write this, to coalesce ideas and narratives I have been teaching now for a decade, and to present a concise and unique view into a complex issue for a range of professionals and students interested in this fascinating subject.

Mexico City, Mexico David Koepsell
November 10, 2016

Chapter 1
Introduction to Scientific Integrity and Research Ethics

Abstract The institutions of science have been studied methodically for some time. We know that for the general, steady progress of scientific discovery to continue, certain social-institutional norms must be adhered to. Robert Merton called these norms part of the "ethos" of science. Science and its component institutions must operate according to the norms of universalism, communalism, disinterestedness (equipoise), and organized skepticism. When scientists or others working in scientific institutions ignore or oppose these norms, science may become pathological and its progress may slow or halt. In this chapter I discuss those norms, their meanings, and provide some examples for their correct and pathological functioning. As well, I provide a brief background to the development of modern scientific ethics.

1.1 What Is "Science" and Its Ethos?

At the heart of the discussion in the chapters to come of the ethics required by science and its institutions is some understanding of what science is. As a general frame for our discussions below, I will focus on a view of science that is not without critics, but one which I think matches closely the nature of science as seen by the majority of its practitioners, whether consciously or not. Science is an institution that proceeds best when scientists work according to certain principles which are not necessarily "ethical" principles, but which demand certain behaviors that we might well call ethical within the domain of research. Specifically, for science to work properly, scientists must embrace the principles of communalism, universalism, organized skepticism, and disinterestedness. Robert Merton first identified these elements of the "ethos" of science in his work as an ethnographer of science, and he noted that failures to abide by these principles may result in scientific research programs' failures.

Science must be universal for research programs to succeed or indeed have any meaning. The truth must not be specific to any one culture, time, or place, but rather inherent somehow in nature and discoverable by the methods of science. It is a communal endeavor, pursued by various people at various times, observing, making hypotheses, testing, and devising theories all in reference to the work of others. No theory is ever the end of science either, and even as confidence builds in any one scientific theory, it is always contingent. Thus, scientists must remain skeptical, as individuals and as groups engaged in various research programs, always willing to set aside a beloved theory in the face of new evidence or falsification. Finally, scientists must not be vested in a particular outcome, and must attempt only to seek the truth, regardless of how their research turns out. Most science leads to dead ends and produces no great breakthrough, but it is on the back of such research that discoveries occur. Disinterested scientists pursue the truth no matter where their studies lead them, disinterested in achieving a particular, sought-for outcome.

When scientists working in its various institutions do not individually and collectively abide by its ethos, then science may go awry. Research may not lead to the truth, delays in advances may be suffered, or setbacks may occur that have repercussions throughout an entire research program and across borders. In the worst situations, people may be unnecessarily harmed. We will see some examples of all of these in the pages to come. In the following chapters we will examine, in light of the ethos of science, various ethical duties of scientists, their hosts, and their funders, and look at the developments of modern codes of conduct and ethics as they relate to science and its methods. Meanwhile, let's examine how the notion of "ethics" in scientific research became such a concern for so many, and then look at how scientific integrity, or research ethics, emerges from both the ethos of science and a history of errors and harms in the development and pursuit of modern science.

1.2 Early Lapses in Science and Ethics

Science was not always pursued in ways that conform with our modern notions of scientific ethics, especially with regard to the use of human subjects. It is from a rather sordid history of scientists' use of humans as subjects that the modern version of bioethics and its various related applied ethical fields evolved. One glaring example is the case of Edward Jenner. We should all be quite thankful for Jenner's discovery of the relation between cows, milk maids, and smallpox because it led to the entire modern practice of vaccine development, and saved literally millions of lives and avoided countless amounts of human suffering. In the late 1700s, Jenner noticed that milkmaids, who came into daily contact with milk cows through milking them by hand, seemed to suffer fewer cases of smallpox, which was a particularly gruesome and virulent affliction back then, taking millions of human lives over the course of human history, and leaving countless others permanently scarred.

Jenner hypothesized that cowpox, which appeared in many outward respects similar to smallpox but which afflicted cows as opposed to humans, somehow conferred upon those who are exposed to it some resistance to smallpox. The hypothesis became the basis for modern vaccine theory and the development of vaccines that have since helped to eliminate some diseases, including smallpox, from human populations, and save and improve countless millions of lives. Unfortunately, because of the lack of any notion of scientific ethics in the use of human subjects, Jenner's experiments proving his hypothesis were hopelessly unethical.

To test his hypothesis, and create a means of preventing non-milkmaids from developing smallpox, Jenner scraped the pus out of cowpox pustules and then exposed subjects who had never contracted smallpox with the cowpox pus. To then test the ability of the cowpox exposure to help prevent smallpox, he then needed to expose his subjects to smallpox. Famously, he chose as a subject of this experiment the eight year old son of his gardener, a boy named James Phipps. More specifically, he took the pus from the hands of a milkmaid who has contracted cowpox (humans who get it get pustules but rarely any other symptoms, like chickenpox but much less painful), and inoculated young James Phipps with the cowpox pus. Then he exposed Phipps to smallpox. Phipps was one of seventeen subjects of this experiment, and like the others failed to contract full-blown smallpox, very fortunately proving his hypothesis correct that cowpox exposure produces a partial immunity to smallpox. In fact, the term "vaccination" derives from the Latin word for cow – *vacca*. Of course, the entire experiment was handled quite unethically compared to modern standards of using human subjects. Thanks, however, to this incredible discovery, and a concerted effort on behalf of medical scientists worldwide, smallpox was declared extinct from the natural environment by the World Health Organization in 1979, the first such disease to be so eradicated.

Jenner's science was unethical for a number of reasons that now seem quite obvious to us, including his use of an underage subject who was presumably unable to properly consent to the procedure, and who was then exposed to a deadly pathogen without the benefit of previous animal experiments. Nonetheless, without any general consensus about using humans as subjects in experiments, and the restrictions about how one should do so, experiments like Jenner's continued, and worsened over time. Notably, the world's attention to the problems of using humans as scientific subjects without consent and other protections became particularly focused after World War II following the Nuremberg trials.

1.3 Nuremberg and Its Progeny

At the conclusion of World War II, the allies established tribunals to bring war criminals to justice, although for some of the crimes alleged there were no international rules, treaties, or laws by which to charge and convict. Nonetheless, in some of these cases the tribunals found that unwritten laws governing proper moral conduct, even in the case of war, required conviction for "crimes against humanity." The modern regime of international human rights and moral and ethical duties derives from much of what occurred at Nuremberg.

One of the tribunals convened at Nuremberg presided over the "doctors' trial" in which a number of Nazi medical professionals were prosecuted for crimes against humanity in the conduct of scientific research, much of which used prisoners of concentration camps. At the doctors' trial, 23 defendant Nazi medical doctors were tried for crimes against humanity in their use of human subjects for experiments, some of which were deadly, and others permanently disfiguring. The doctors' defenses included claims of humane euthanasia, and the lack of any governing standards for experimentation on human subjects, especially the terminally ill, and sometimes involving studies that yielded important scientific results.

The tribunal at Nuremberg had no code or law with which to render a verdict, but rather appealed to general principles of morality, finding in so doing that there are moral boundaries to science that must guide its proper conduct, especially when using human subjects. Although scientists had since even before Jenner's experiments conducted experiments without any particular ethical constraints, the Nuremberg court found that indeed scientists ought to act within the bounds of certain constraints. Their decision forms the basis for modern applied ethics, especially as applied to human subjects experiments.

The Nuremberg Code, as it has come to be known, is described as a series of duties owed by scientists to human subjects and to society, and has become the basis for internationally-recognized boundaries of behaviors in conducting scientific study. The code describes ten specific duties:

1. Voluntary, well-informed, understanding consent of the human subject in a full legal capacity.
2. The experiment should aim at positive results for society that cannot be procured in some other way.
3. It should be based on previous knowledge (like, an expectation derived from animal experiments) that justifies the experiment.
4. The experiment should be set up in a way that avoids unnecessary physical and mental suffering and injuries.
5. It should not be conducted when there is any reason to believe that it implies a risk of death or disabling injury.
6. The risks of the experiment should be in proportion to (that is, not exceed) the expected humanitarian benefits.
7. Preparations and facilities must be provided that adequately protect the subjects against the experiment's risks.
8. The staff who conduct or take part in the experiment must be fully trained and scientifically qualified.
9. The human subjects must be free to immediately quit the experiment at any point when they feel physically or mentally unable to go on.
10. Likewise, the medical staff must stop the experiment at any point when they observe that continuation would be dangerous.

Although based upon this previously unwritten code, verdicts of guilty for crimes against humanity were handed down in the Doctors' Trial, it would still be decades before these duties became formally codified through laws, rules, or regulations. Before describing the history between the Nuremberg Code and the development of modern codes of research ethics, we should look at the duties described by the code and see whether and how they may emerge from the study of ethics and morality as it had been conducted for millennia, because the principles expressed do not emerge out of thin air. They are clearly derived from the philosophical study of ethics, and the notion that the Nuremberg tribunal found moral principles applicable to scientists generally and timelessly, rather than created them *sua sponte*, is important for the future as new social and technical challenges push the limits of our scientific study and test our presuppositions about what is ethical.

1.4 The Origins of the Nuremberg Principles

For the past 2000 years or so, philosophers have been trying to describe the nature of "the good." Today, there are perhaps three major schools of ethical theory that have emerged through philosophical study, though there are a number of variants of each and philosophers who debate the various classifications. For our purposes, we will focus on the three major ones without getting into the finer points of debating whether there are others, or whether they are properly classified, nor shall we concern ourselves deeply with various versions of each. The three in chronological order of their development are: virtue ethics, deontology, and consequentialism. Each of the major theories is in fact represented in the Nuremberg principles, as we shall see. In discussing each, we will also limit ourselves to the most famous proponents of each of these broad theories, including names that are famous even outside the field of philosophy, including Aristotle, Immanuel Kant, and Jeremy Bentham. Together, they represent the foundations for much of the modern study of ethics, and their theories inform ethical decision-making in nearly every instance of applied ethics, although there have been some additions, alternations, and refinement to the original theories over time. They are also the theoretical bases for the Nuremberg Principles.

1.4.1 Virtue Ethics

For the Greeks, notably Plato and then more fully expressed by Aristotle, the basis of the good was in cultivating a good character, which involved the development of certain virtues. Plato described four "cardinal" virtues: prudence, justice, fortitude, and temperance, development of which was necessary through proper study and the moderation of our emotions through the development of our facilities for reason. Aristotle's Nicomachean Ethics refined a view of virtue ethics by attempting to bring a measure to the notion of a virtue, and by expanding upon the list of virtues. Again calling upon our facilities for reason, Aristotle viewed the virtues as a mechanism by which we attain the good life, bettering our capacities for honesty, pride, friendliness, wittiness, rationality in judgment; mutually beneficial friendships and scientific knowledge.

For the Ancient Greeks, the proper aim of life is *eudaimonia*, or roughly "the good life," "happiness," or "well being." Through the development of good habits of character, we can achieve eudaimonia and it is moral in the sense that it is actively sought rather than blindly. Our goal should be to seek the good life through the active contemplation of the virtues and attainment through the cultivation of good habits of character by study and reflection. While virtue ethics offers us no specific guide for decision-making, virtue ethicists often hold that the measure of the good is not in the action but in the person, and that a virtuous person properly educated will make judgments that generally promote eudaimonia.

Aristotle's refinement of virtue ethics is in part his formulation of the relations of the virtues to one another and to "vices." What is not virtuous is "vicious," and thus the term we have for "vice." For Aristotle, the virtues sit in the mean between two vices, they are not the polar opposites of the vices. Rather, for instance, the virtue "courage" sits in between cowardice and rashness at two extremes. Our tendencies toward the vices are driven by two different types of emotion, and it is through the rational control of those emotions, through the studied and habitual cultivation of our characters, we can keep those emotions in check, be virtuous, and thus best achieve eudaimonia.

A virtuous scientist would, presumably, act in accordance with the principles expressed in the Nuremberg Code because those principles are rational, she would be honest, just, temperate, and prudent. It is also worth noting that Hippocrates developed the famous Hippocratic Oath, which expresses the nature of a virtuous physician, around the time Plato and Aristotle were expressing a notion that we call virtue-based ethical theory. Virtue ethics remained a dominant ethical theory for nearly a thousand years, adopted by Christian scholastics, notably discussed by St. Thomas Aquinas and expanded to include such "Christian" virtues as faith, hope, and charity. As such, virtue ethics is deeply ingrained in western thought even today, and the language of virtue ethics is familiar to us as we continue to regard certain traits of character as better than others, and indeed we call them *virtuous* even when we are not fully versed in the theory of virtue ethics.

1.4.2 Deontology

Another approach to ethics is based upon the notion of duties. The term "deontology" derives from the Greek for duty – *deon.* Deontology overcomes a major limitation of virtue ethics theory in that it is meant to provide some guide for action as opposed to individual character. According to deontological ethics, we must abide by certain duties which can be discovered through a number of means. In rights-based deontology, our duties stem from our obligation to recognize and protect various rights (like life, liberty, property, etc., according to philosopher John Locke). Locke, for instance, views humans as having been endowed naturally with certain rights, and from those rights flow obligations or duties to abide by the rights of others. In this discussion, we will focus on the deontology of Immanuel Kant, who devised what he thought to be a scientific argument for the nature of duties that apply to everyone universally, and without regard to any divine source. Kant's deontology endures in the Nuremberg Code and elsewhere, and is perhaps the most prominent example of a fully developed deontological ethics.

1.4 The Origins of the Nuremberg Principles

Kant defines the only good as that which derives from what he calls the "good will" which must come from a sense or moral duty, as opposed to any particular desired outcome from an action. Rather, we must be motivated by our equal, inherent dignity toward what he calls the "categorical imperative" which is the duty we owe universally and without qualification. Kant formulated his categorical imperative in at least three different ways throughout several works, including the following:

- Act only in such a way that you would want your actions to become a **universal law**, applicable to **everyone** in a **similar situation**.
- Act in such a way that you always treat **humanity** (whether oneself or other), as both the **means** of an action, but also as an **end**.
- Act as though you were a law-making member (and also the king) of a hypothetical **"kingdom of ends"**, and therefore only in such a way that would **harmonize** with such a kingdom if those laws were binding on all others.

The categorical imperative in practice requires us to do at least two things: (1) treat others as though they are "ends in themselves" rather than as instrumentalities, deserving of equal inherent dignity with us, and (2) not do anything if we cannot universalize it without what Kant calls "contradiction."

We can see in various parts of the Nuremberg Code some reference to deontological ethics, specifically regarding notions of justice, not using humans as means to ends, and in notions of autonomy. Natural-rights or Kantian deontology is at work behind many of our current ethical and political institutions, and influences applied ethical decision-making, especially in bioethics. Critical to its distinction from consequentialism is the categorical nature of rights and duties, and the rejection of a means-ends analysis to achieve the good. Consequentialism does the opposite.

1.4.3 Consequentialism/Utilitarianism

The English philosopher Jeremy Bentham also sought an ethics based not in some divine command but rather in empirical reality. He concluded that desire for pleasure and avoidance of pain are universal, and thus these two universal, empirical values are the measures for ethical behaviors. According to Bentham, we can determine the most ethical action by employing a "hedonic calculus" and determining what will promote the greatest pleasure and least pain. Because pleasure and pain are both so universally desired and reviled respectively, their basis as an objective measure of value should be clear, according to Bentham, and can be used as the foundation for "the good."

Bentham's hedonic calculus considers the total quantity of pleasure in the world vs. that of pain, and suggests that we ought to choose the action which increases overall pleasure (the net good) in comparison with the net pain produced. A rough restatement might be: do that which produces the greatest pleasure and avoids the most pain. The hedonic calculus offers us, according to Bingham, a means to judge the good without resort to unfounded assumptions, such as those we make about the nature and existence of virtues or duties. There have been some refinements to consequentialism, and it comes in various forms.

John Stuart Mill, Bentham's student, refines consequentialist "Utilitarianism" by offering gradations of pleasures. The base, bodily pleasures are ranked below the aesthetic and intellectual by Mill's form of utilitarianism. Moreover, there are act and rule-utlitarians. Act utilitarians consider the hedonic calculus on a case-by-case basis, measuring the net utility of each choice we make, and rule-utilitarians gauge the overall societal utility of adopting certain rules. The reason for one choosing rule-utilitarianism over act- is clear if we consider a classic conundrum raised by act-utilitarianism. Below we will consider some of the objections and problems with each of these theories, and then finally look at how they have been actually adopted in various parts of the Nuremberg code and its progeny that enshrine various principles of bioethics.

1.5 Some Problems with Ethical Theories

Each of the major ethical theories described briefly above has a list of objections, and this is one reason there is no general consensus about which, if any, is best. Virtue theory, for instance, offers us no guide to action. It focuses on the development of individual character, by the cultivation of various "virtues" and the naming of the virtues itself is poorly founded. One might well chose another set of virtues than did Plato, as Aristotle did, according to one's own whims. Having chosen some arbitrary set of virtues, moreover, provides little-to-no guidance for how to behave, what choices to make, given an ethical dilemma. For applied-ethics purposes, while one might well support the general principle of education to improve understanding and understanding of the virtues (once one settles upon what they are or should be) to improve one's ethics in general for a profession, virtue ethics does not help with individual cases. It doesn't help us arrive at an answer for a specific problem.

Deontology too suffers some shortcomings, including a major problem with a lack of hierarchy in Kant's duties. If one has duties that conflict, how does one choose which to pursue? An example that is generally used is a potential conflict between the duty to not lie (which Kant's view concludes is absolute) and the duty to protect the life of another. So, if the Nazi Gestapo comes knocking on your door because you're hiding a Jew who would be taken away to a concentration camp, and you are asked if the person is there, Kant and others who view duties as categorical and non-hierarchical would say you must tell the truth. This is of course a rather unsatisfying answer, and runs counter to many of our ethical intuitions. Surely we can bend or break some duties to abide by others according to some hierarchy of duties. Some recent deontology considers and proposes such alternatives. A more foundational problem of deontology is the leap required in order to accept that some duty exists. As with virtues, we must at some point blindly accept a duty's existence without requiring that it be based on something empirical.

Consequentialism offers a number of problems as well. The notion of the greatest good (happiness, utility) for the greatest number appears to most at first blush not only appealing but indeed grounded in something universally valued: pleasure. Yet on analysis, the hedonic calculus in both the act- and rule- forms leads to unsatisfactory examples. If an act, for instance, sacrifices just the right amount of people's happiness for the increased pleasure of more people, it would be ethically acceptable. As long as the hedonic calculus leads to a net increase of pleasure, we should prefer it. This is so regardless of the act. Thus, enslaving just the right number of people would be acceptable as long as net happiness is increased, despite the loss of freedom by the enslaved populace. Indeed, any number of horrors might be tolerated if conducted on a sufficiently-sized portion of the populace so long as net happiness is increased. Rule-utilitarians might try to avoid the unsatisfactory results of act-utilitarianism by arguing that adopting a rule that, for instance, prohibits slavery helps best to increase net happiness in the long run, yet the rules created by this form of utilitarianism too pose a problem.

Consider the *Les Miserables* problem. Jean Valjean in Victor Hugo's masterwork suffers for decades in prison as the result of the theft of a loaf of bread. Act-utilitarianism might well have admitted the morality of the theft as his family was starving, and the minor loss of happiness to the baker would be offset by the ability to live by Jean Valjean's family. However, the general rule that one might posit against theft could well be supported by the overall increase of utility which results from rules against theft. Rule-utilitarianism suffers similar problems in the failure to admit of exceptions as does deontology. It is entirely possible that ethical theory fails because there is no such thing as "the good' in a metaphysical sense. In other words, ethics may be created by humans, there may be no foundational basis for "the good" and we could all be engaged in simply stating preferences when we "do" ethics.

1.6 The Modern Bioethics Regime

Even after the Nuremberg trials and the introduction of the "code" through the Doctors' trial, it took a while before science began to form institutionalized measures to help ensure its application through science. Meanwhile, numerous ethical lapses continued, some of which were on similar scales to what was revealed at Nuremberg.

During the Cold War, a string of incidents proved that scientists still had not perfected their abilities to restrain themselves and to conduct scientific studies using humans in ways consistent with the Nuremberg principles. During and after World War II as the Cold War was beginning and the two major superpowers were racing to develop more nuclear weapons, for instance, the governments were very interested in the effects of radiation on humans. The US government conducted a number of troubling, secret experiments. In 1995, over 1 million pages of previously classified documents were released by the US government, some of which detailed experiments such as: giving radioactive materials to disabled children without their knowledge or consent, similarly exposing US soldiers to highly radioactive materials, secretly exposing prisoners' testicles to highly radioactive materials resulting in horrifying birth defects, exposing US citizens who checked into hospitals to radioactive materials without their knowledge or consent, and numerous other such experiments all conducted outside of the constraints of Nuremberg principles.

In the 1950s and 1960s, also in the context of the Cold War, the two superpowers were conducting experiments in mind control, also in flagrant violation of the Nuremberg norms. In the US, the so-called MK-Ultra experiments involved among other things the surreptitious dosing of innocent and unknowing civilians with various psychotropic agents, including LSD. Again, it was only years later that the details of these experiments were revealed and the full extent of their damage to unsuspecting and unwilling participants is still not known, although we know that at least one subject died as a result of the study.

Finally, non-military studies that also mistreated human subjects according to the Nuremberg principles also were ongoing well into the 1960s and 1970s. One fascinating study that reveals quite a bit about how scientists themselves may, despite their best intentions, do morally questionable things, is one done by the famous Stanley Milgram. Milgram's most famous experiment involved recruiting volunteers who were told they were going to assist in a study about learning. The volunteers would be told by the experimenter to push a button that would administer a shock to what they were told was a test subject, but who in fact was an actor. No real shocks were administered, but the real subjects of the study, the volunteers who were told they were administering shocks, did not know that. During the course of an hour session, the scientists would tell the volunteers to give shocks to the subject even as the supposed voltage was increased, eventually into dangerous territories. Almost invariably, the volunteers continued giving the shocks even when the actors who were pretending to be the subjects feigned serious injury, and in at least one case, a heart attack. The study revealed something rather important about the human character, something which subsequent studies by Milgram and other have confirmed: people will listen to authority and do as it says even when it contradicts their own conscience.

Milgram's study was of course also unethical as it did not allow the real subjects to have properly informed consent. Similar studies have since been conducted, in which the subjects were given proper information and consented accordingly, and which have helped reveal the same information about authority and conscience. The initial study would not be approved today, and subjects of the study did indeed suffer as a result, in some cases requiring therapy for years for their post-traumatic stress as a result of finding out they were capable of hurting people on the word of some authority.

Finally, the Tuskegee syphilis study involved behaviors egregious enough, at just the right moment in US civil rights history, to permanently alter the institutional and legal landscape of human subjects research. Using mostly poor, black, and often illiterate share-croppers in the US south, the study conducted over 40 years followed the course of degeneration of those suffering from syphilis. The physicians involved told the subjects they were getting free health care from the US government as the experimenters followed and published papers regarding the ongoing, degenerative effects of syphilis. 15 years into the study, it had been discovered that penicillin was an effective cure for syphilis, but the more than 600 subjects of the study were not informed nor treated with it during the remaining 25 years of the study. It was only the 1970s, with information leaked to the press by a whistleblower, that public attention and indignation properly came to the physicians and governmental organizations involved and the study was shut down and investigated.

It was the public attention to the Tuskegee study and its revelations that led to a governmental review, and the eventual publication of the Belmont Report, which expressed the principles first enunciated in the Nuremberg opinion, and which formed the ideological basis for the creation of official institutions to oversee research on human subjects, namely: the creation of the Office of Human Research Protections, the creation of ethics committees to pursue those goals, and the legal and regulatory apparatus that now oversees and provides ethical evaluation and oversight to all human and now animal studies. Although the Helsinki Declaration, which is an international statement also concerning experiments on human subjects, was first developed in the 1960s, following Tuskegee it was significantly revised and amended and the US institution of ethics committees included.

The modern environment for research ethics has developed out of this history, and the current institutional arrangements both within and among nations is heavily influenced by the ongoing problem, and the very public concern as a result, of scientists' apparent ongoing failures to abide by principles enunciated following the atrocities of World War II. Even now, there is little in the way of conformity in approaching the problem of scientific integrity and research ethics among universities, research centers, and in ongoing and now often complex multicenter studies and funding programs. Understanding why scientists might do unethical things has proven to be more fruitful than teaching them not to. The current regime of oversight and regulation is burdensome and often disapproved-of by those who must now navigate more paperwork and bureaucracy. Yet every year, it seems, new ethical failures of both academic and scientific integrity, sometimes still involving human subjects, come to light. The overall result of each of these instances is erosion of public confidence, which sometimes has real, budgetary repercussions, and setbacks in the worst cases to science in general, as well as to particular fields of study.

In the following chapters, we will explore these issues together, offer some means to help educate and prevent such lapses, and hopefully help underscore the reasons inherent both in the ethos of science, and in our appreciation of science itself as a vital human institution, for avoiding the harms and wrongs described below.

Study and Discussion Questions

1. How was Jenner's experiment on smallpox unethical? How could it have been redesigned to make it ethical in relation to modern bioethical principles?
2. What are the ethical theories that help inform the Nuremberg Code? Which principles are based upon which philosophies?
3. How were the Tuskeegee, MK-Ultra, and Radiation studies unethical? How could they have been revised to be ethical in relation to the Nuremberg Code?
4. Is the Helsinki Declaration law? Is it binding upon anyone? Why or why not? What purpose does it serve, in your opinion?

References

Alexander, Larry, and Michael Moore. 2007. *Deontological ethics*.
Annas, George J., and Michael A. Grodin. 1992. *The Nazi doctors and the Nuremberg Code Human rights in human experimentation*. New York: Oxford University Press.
Baker, Robert. 2001. Bioethics and human rights: A historical perspective. *Cambridge Quarterly of Healthcare Ethics* 10(03): 241–252.
Bentham, Jeremy. 2000. *Selected writings on utilitarianism*. Hertfordshire: Wordsworth Editions.
Boire, Richard G. 2001. On cognitive liberty. *The Journal of Cognitive Liberties* 2(1): 7–22.
Book, I. 1980. *Aristotle, Nicomachean ethics*.
Cave, Emma, and Søren Holm. 2003. Milgram and Tuskegee—Paradigm research projects in bioethics. *Health Care Analysis* 11(1): 27–40.
Faden, Ruth R., Susan E. Lederer, and Jonathan D. Moreno. 1996. US medical researchers, the Nuremberg Doctors Trial, and the Nuremberg Code: A review of findings of the Advisory Committee on human radiation experiments. *JAMA* 276(20): 1667–1671.
Hugo, Victor. 1987. *Les misérables*. Trans. Charles E. Wilbour, 16. New York: Modern Library, 1992.
Hursthouse, Rosalind. 1999. *On virtue ethics*. Oxford: Oxford University Press.
Johnson, Robert. 2008. Kant's moral philosophy. *Stanford Encyclopedia of Philosophy*.
Kant, Immanuel. 1997. Groundwork of the metaphysics of morals (1785). In *Immanuel Kant: Practical philosophy*, vol. 80. Cambridge: Cambridge University Press.
Lee, Martin A., and Bruce Shlain. 1992. *Acid dreams: The complete social history of LSD: The CIA, the sixties, and beyond*. New York: Grove Press.
Merton, Robert K. 1942. Note on Science and Democracy, A. *Journal of Legal and Political Sociology* 1: 115.
Mill, John Stuart. 1987. *Utilitarianism and other essays*. New York: Penguin Books.
Riedel, Stefan. 2005. Edward Jenner and the history of smallpox and vaccination. *Proceedings (Baylor University. Medical Center)* 18(1): 21.
Weindling, Paul. 2004. *Nazi medicine and the Nuremberg trials: From medical war crimes to informed consent*. New York: Palgrave Macmillan.

Chapter 2
Research (Mis)Conduct

Abstract It seems that increasingly scientific publications are being pulled from publications following discovery of some fraud, misrepresentation, or other misconduct related to the science or the data involved. The manipulation, misrepresentation, and fraudulent use of experimental results has been a problem for science since science began. Whether it is increasing in frequency, or rather just becoming noticed and discovered more thanks to growing awareness and vigilance, would be an interesting subject for study. Here, we will examine its nature, its various forms, some of its causes, and ways to find and prevent scientific misconduct of various kinds, specifically those we are apt to call "fraud."

2.1 Sources and Citations

A fair segment of research misconduct comes from failures to properly acknowledge where data comes from, as benign as that may seem. Providing proper means for others to trace back the sources of scientific data, and thus to help them to reproduce (or contradict) results, is at stake, and is part of the ethos of science itself. Because science is a communal activity, depending upon a community of researchers doing basic research and challenging it over time, scientists must provide for others in their community the means to check and challenge, and hopefully confirm the results of experiments. Failures to properly attribute the sources of data, or fraudulent manipulation of the data, may result in harm to the scientific community as well as the public upon which it depends. Even minor lapses that prevent the full assessment of the data used may make progress difficult in a field, even where the lapse may at first appear to be innocent.

The nature of the scientific enterprise demands that observations and experiments be either verified through independent researchers, or falsified. In order to do so in a way that is scientifically valuable, all relevant features of an observation or experiment must be captured, if, for instance, we are to avoid the sort of pathological science that can result otherwise. When we speak of "pathological" science, we mean that science which isn't pursued according to the norms of science, and which often ends up being harmful to scientific progress and to society as well. Sometimes, in the pursuit of a particular hypothesis or theory, scientists fail to be properly disinterested, and may overlook some data, may even "massage" it in a way that makes it conform better to their view of how the experiment should turn out, or at the worst may manipulate the data consciously, misrepresent it non-innocently, and attempt to fool the rest of the scientific community for whatever reasons.

The most important part of a scientific study is the data, although the publication and dissemination of the results are what tend to focus most of our attention both as scientists and as members of the public. Positive results also tend to be most interesting, even though negative results as equally important for the progress of science. The scientific community must have access to the data that is behind published results if science is to proceed non-pathologically. In order to replicate an experiment, generally the raw data must be somehow well curated and made available for other scientists to examine if necessary. This may become necessary when attempts to replicate an experiment based upon published results run into difficulties. One ancient example of such a failure, with even potentially fatal results, is the famous star map, the *Almagest* of Ptolemy.

For centuries, Ptolemy's star map became the essential source of data concerning the visible stars, and other maps and sources more or less disappeared from sight and opportunity for scrutiny. Ptolemy's was the definitive star map. Ptolemy was a Greek academic who did much great work in bringing together as well as making empirical observation of data regarding the Earth and the solar system, as well as basic work in optics and other fields. He lived from A.D. 90 to 168, doing most of his observations, especially those connected with the *Almagest*, in Alexandria, Greece. Hipparchus was a lesser-known scientist working in Rhodes, which lies to the south of Alexandria by about 5 degrees. Hipparchus lived from about 190 B.C. to 120 B.C. He had apparently made detailed observations of the stars as well, although his original observations were not preserved, though Ptolemy and others made reference to them. Both Hipparchus and Ptolemy also apparently made use of Babylonian star observations, and referenced them, although those original observations also do not survive.

The appearance of the stars in the sky changes over time, and this was known even by Ptolemy and others in the ancient world. The movements of the stars over time is based upon a variety of movements combined, including those of the stars through the galaxy, and the procession of the orbit and rotation of the Earth. Ptolemy mentions Hipparchus's work, but claims that the data regarding the stars referenced in the *Almagest* are from his own observations. There are a few reasons why, over time, numerous scientists have come to doubt that this could be so. Isaac Newton was among those who have accused Ptolemy of, essentially, plagiarism in the modern scientific era. Noting a number of factors that make it more likely than not that Ptolemy used some or even much of Hipparchus's data rather than collecting, as he claimed, all of his own, the debate continues today about just how much of the data in the *Almagest* comes from other sources. One recent statistical study, relying on the observations of the southernmost stars in the catalogue, puts the probability that the stars noted were observed from Rhodes, where Hipparchus did his observations, at 90%, as opposed to about a 10% probability that the same stars were observed from Alexandria, where Ptolemy did all his work and appears to have lived his whole life.

2.1 Sources and Citations

Of course, the scientific standards for citations and reference have changed a lot since ancient Greece, but the case of Ptolemy raises some interesting questions about the source and impact of the duties of scientists to properly reference the sources of their data. Without a clear path for scientists to trace back the proper sources for each of Ptolemy's reported star observations, his observations are more or less useless to future scientists, despite their contemporary use and accuracy. This is because, as mentioned above, the movements of the stars through the sky over time are both enormously interesting and incredibly useful. Consider that until the 19th century, most navigation of ships depended upon celestial observations, and the accurate prediction of star locations projected into the future was absolutely essential to making useful celestial maps for navigation. Given the five degree difference between where Hipparchus likely made many of the observations in Ptolemy's maps, and the difference in time between when Hipparchus would have made those observations and when Ptolemy claims to, the error in the locations based on the alleged times of observations would have, if depended upon without corroboration, resulted in some rather significant and potentially disastrous discrepancies for navigation by the stars. Some have noted that Ptolemy appears to have tried to correct for errors by either adjusting for precession based on time, or have made significant errors in his own observations that make it appear that he attempted to cover for the anticipated errors. In any case, because of his failure to account for these adjustments, or to properly cite the sources for much of his data if as seems likely he did not do many of the primary observations, his *Almagest* becomes more or less useless for science, as well as dangerous for a certain portion of the public.

A clear data trail is not only good manners, providing acknowledgment of the source of data and work of those responsible, but also the means by which to check and to refine hypotheses and theories over time based upon observations made in the past. Failing to leave such a trail does a disservice to other scientists, and runs afoul of the ethos of science which regards its institutions as necessarily communal. Because scientists work in a community of researchers, and the interaction of that community by observation, challenges, and refinement of theory over time from one research group to another all depend upon accurate accounting for the means, place, and time of each observation. Because, for instance, Ptolemy didn't give the sources for at least some of his alleged observations, it is impossible for researchers to trust their accuracy and use them for improving the corpus of knowledge about the stars. Ptolemy's *Almagest* was used for nearly 1000 years as the standard for celestial data, and for navigation. Its trustworthiness became taken for granted, and in that time there was very little in the way of competing science, new observations, or challenges to its accuracy. Moreover, the data that Hipparchus and others had gathered about stars has been lost in the meantime, its importance to the body of knowledge about the sky forgotten due to trust in one accurate, primary source that has turned out not to be so primary. This is a tremendous loss for science, and may have set the field back for centuries as opposed to provoking further and ongoing observation and data collection.

2.2 Data Collection, Manipulation, and Curation

Another more recent example is that of Robert Millikan and his attempts to measure the charge of an electron. At the turn of the twentieth century, there was some doubt as to the nature of the charge of an electron, with some claiming that it was graded, coming in a range of values, and others arguing that it was unitary. Millikan believed that the charge of the electron was unitary, while Felix Ehrenhaft believed that it came in degrees. Both designed quite delicate, slightly differing experimental mechanisms to attempt to measure the charges and came up with different conclusions. The problem presented to both in designing their experiments was that the charge of the electron is so small, and the mechanisms for measuring it at the time were as yet still quite primitive for the task. The net result of both of their experiments was that each found charges that were more or less all over the place, with certain clusters but large groups of outliers, making it difficult to come to the conclusion that the electron had a single charge. Except that Millikan did come to that conclusion, and published that conclusion, then received the Nobel Prize a few years later as a result of his science. Based on what we know today, however, his experimental data did not necessarily support his conclusion.

In 1978, Millikan's lab notebooks were discovered and they are rather revealing. Ehrenhaft's notebooks were lost when he fled Austria at the outset of World War II. What Millikan's notebooks reveal is that he didn't keep, much less publish all of the results of his experiments. According to Millikan's notebooks, he ran his experiment involving falling drops of oil a total of 140 times, yet in his published paper, that was to become the basis for his Nobel prize, he reports only 58 drops, stating:

> It will be seen from Figs. 2 and 3 that there is but one drop in the 58 whose departure from the line amounts to as much as 0.5 percent. It is to be remarked, too, that this is not a selected group of drops but represents all of the drops experimented upon during 60 consecutive days ... (Millikan, 1913, p. 138, original italics)

Meanwhile, in his notebook often in marginalia near drops that appear to have been excluded from his final calculations and publication are statements of this sort: "Error high will not use ... can work this up & probably is ok but point is [?] not important. Will work if have time Aug. 22." And "It was a failed run—, or effectively, no run at all." It seems that on several occasions, when the oil drops were not behaving as Millikan expected, he simply stopped the experiment and started anew. On other occasions, where the oil drops acted in better accordance with his expectations, Millikan made some rather joyous remarks in his notebooks, reminding himself to definitely use those results. All of which is of course troubling, even though Millikan's hypothesis has since been repeatedly confirmed.

Ehrenhaft, on the other hand, was also getting a range of results from his experiments, but his reaction to the data differed from Millikan's. Because Ehrenhaft expected a range of results due to his hypothesis of "sub-electrons" having differing charges, he interpreted what was messy data that resulted from the primitive experimental mechanism intended to measure a very small effect as supporting his hypothesis. Millikan, on the other hand, cleaned up data without informing the world, to make it better conform to his hypothesis. Had he published every result of his experiment, it would have looked more like Ehrenhaft's data, and not have supported his hypothesis to the extent that his published results did. He may also not have gotten the Nobel Prize. Indeed, at the time he was awarded the prize, there was an ongoing and lively debate about Millikan's and others' experimental results, with famous scientists taking sides even as others were having trouble replicating the very delicate experimental setup. Nonetheless, the tide of opinion was turning toward a unitary charge for the electron, and subsequent experiments have so far borne it out. But Millikan failed to abide by the ethos of science, and hid an important data trail that could have helped end the controversy sooner, and with the knowledge we have now about his actual observations, casts his character as a scientist into some doubt.

How did Millikan's deception harm science? After all, years of further study have failed to disprove his hypothesis and the unitary charge of the electron is now well-established theory. He was, apparently, right in all his assumptions and his actions could be seen as cleaning up messy data, using his instincts to avoid the error that the primitive measuring apparatus was unable to correct for. There are a number of reasons to criticize what Millikan did, and ultimately his behavior was harmful to science as many, including Richard Feynman, have concluded. Millikan lied, simply, when he claimed in his publication, the one leading to his Nobel Prize, that he was reporting all his observations. He acted in ways that run counter to the ethos of science. His decisions to discard certain results were based upon his expectation of results that conformed to his hypothesis about the charge of electrons. He failed to be sufficiently disinterested and lacked equipoise. He assumed that certain results were experimental errors, and they may well have been (in fact, probably were, since the setup was so delicate and prone to such errors). But his assumption, and how he acted upon that assumption, robbed the scientific world of important opportunities for inquiry. This violated the ethos of communalism. Millikan's publication, and the certainty with which he presented it, lent error to science for decades, as it turns out.

Millikan's measurement of the charge of the electron was slightly off. As Richard Feynman notes in his essay "Cargo Cult Science," after Millikan, and bolstered by his Nobel Prize on the subject, scientists who measured the charge of the electron more accurately were reluctant to report that data given its disagreement with Millikan's, even though theirs was more accurate. Just as Millikan was expecting a result that would conform to his hypothesis, so subsequent measurements by other scientists were discarded for their failure to meet expectations, as opposed to treating them as data that needed to be better understood. As Feynman so eloquently and plainly put it:

> The first principle is that you must not fool yourself--and you are the easiest person to fool. So you have to be very careful about that. After you've not fooled yourself, it's easy not to fool other scientists. You just have to be honest in a conventional way after that.

Millikan, unfortunately, both fooled himself and, perhaps inadvertently, a generation of scientists following. It would be unfair to characterize Millikan's work as outright fraud. His manipulation of data was not fabrication but cherry-picking:

choosing data that most closely conformed with his hypothesis while running an experiment that was perhaps the best he could design for the task at the time. It seems clear, even from the lab notebooks that are now part of the record of this case, that Millikan did not intend to deceive, but rather to make more clean the case he was making for the unitary charge of the electron with experimental data that he felt best represented it. His failure was not invention, but more like looking at the relation between hypothesis and data through rose-colored glasses, as optimistic about the conformity of nature to his vision of it.

Sometimes researchers go even further than Millikan, and the line between what Millikan and countless other researchers have done, in both fooling themselves and thus the rest of the scientific community and ultimately the public, and outright fakery, may not be as bright as we would wish. Very recently, in the areas of social and behavioral psychology, two rather different but prominent cases illustrate how that line may become blurred, and in the worst cases crossed.

2.3 Correlation Without Causation

Significantly bolstering the rash of modern books of positive psychology (which take off where Norman Vincent Peale's *The Power of Positive Thinking* left off in the 1950s) is a now mythic graph that "shows" that flourishing, successful individuals' ratio of positive thought to negative is roughly 3 to 1 (2.9013, to be more precise). Science backs it up, and with a nice science-ish number and complex mathematics behind it. The paper introducing this magical number was entitled "Positive Affect and the Complex Dynamics of Human Flourishing" and was co-authored by Barbara Fredrickson and Marcial Losada. It has been cited more than a thousand times. It also provided the scientific grounding for a new bevy of positive thinking books and theorists, including such best-sellers as *Flow: The Psychology of Happiness* by Mihaly Csikszentmihalyi, *Authentic Happiness: Using the New Positive Psychology to Realise Your Potential for Lasting Fulfilment* by Martin Seligman, and Fredrickson's *Positivity*. Both Seligman and Csikszentmihalyi had promised to found a new, scientific "positive psychology" on a rigorous, evidence-based foundation in their jointly-written, "Positive Psychology: an Introduction" published in 2000. Unfortunately, the work that has formed far too much of that foundation over the years, has proven to be unsound. There may be something to positive psychology, but the "Losada line," cited thousands of times in works of positive psychology as support for the role of happiness in thriving, is not the foundation we were looking for.

The Frederickson and Losada article purported to show a correlation between the "positivity" of a group of subjects and their worldly success. Among the subjects interviewed were a group of university students, and their positivity was measured through subjective interviews, then compared with objective criteria regarding their success in an academic milieu: their grades. The Frederickson and Losada article cites to a previous Losada article in which the author first argued for a relation of

2.3 Correlation Without Causation

positivity expressed by members of observed meetings, and the success of teams, using mathematical correlations seemingly derived from differential calculus, and corresponding to the famous Lorenz equations used primarily in fluid dynamics. In the critical article with Frederickson, Losada's previous work is cited as support for the thesis regarding the magic ratio of positivity to success, differential equations and reference to the Lorenz equation help inspire confidence in the scientific grounding of their conclusions, and the rest is citation history.

As it turns out, the math was bunk, and the ratio claimed is not grounded in any sound science. In the paper The Complex Dynamics of Wishful Thinking by Nicholas J. L. Brown, Alan D. Sokal, and Harris L. Friedman, Fredrickson and Losada's ratio is eviscerated, and even Frederickson has now conceded this point, still putting a positive spin on things and clinging to her thesis, despite the lack of solid evidence. The manner in which the "correlation" of observed behaviors, subjective determinations of "positivity", and success is made is more or less woven from whole cloth. Brown, Sokal, and Friedman demonstrate that *even if* there are some arbitrary choices of measurement used in the foundational work (the earlier work of Losada invoking differential equations to try to measure positivity in language use and outcomes), those measurements are not spelled out, and the choices for marking the boundaries of those arbitrary measures are not explicit. There is no way to duplicate the research, and the nature of both subjective and objective measurements of the values involved (positivity, success) necessarily require arbitrary choices in measurement. They are entirely inappropriate for application of differential calculus, and we could achieve more or less any desired outcome by subtly shifting the boundaries of our values in measurement, which is apparently what the authors did to achieve their magic ratio. We should be quite skeptical. Correlation, even if there were one, is not causation. There is simply no correlation, though, and the reams of pages that have been printed, seminars attended, self-help courses taught, and other purveyors of happiness as the cure listened to, have no right to appeal to what became a seminal work in the happiness industry.

The Frederickson and Losada case is notable for the way that, again, data and mathematics are used by the researchers to help confirm something without proper regard for the manner in which the data is measured, correlated, and then represented. Having perhaps already reached a conclusion about some relationship between happiness and thriving, the researchers appear to have used some spurious manners of gathering, measuring, and then correlating their data, as well as an unwarranted an unsupported connection between the data they gathered and an equation best known from fluid dynamics. The net result has been that thousands of scientists have used their paper as the basis for further conclusions that, like in the case of the alleged existence, and subsequent dismissal of N-rays (as we shall see), were never warranted by the basic research. As with the Millikan case, this is not clearly fraud, and the researchers may have successfully fooled themselves primarily without the desire to fool others, but fully convinced of the truth and value of their research. The Brown, Sokal, and Friedman article makes clear the methodological errors that Frederickson and Losada employed in their study, and should serve as a warning to future researchers who too casually try to correlate such complex data with expected effects using what are impressive, but ultimately poorly connected mathematical equations. It is worth noting that Sokal made a name for himself in the study of bad research quite independently of this case.

2.4 Publication Bias as Part of the Problem

One driver of some of the misconduct described above is the desire to publish, to be first, to make an impact. Part of the reason this driver may lead to errors, hasty generalizations, or even fraud as we shall see is the effect called "publication bias" which lies on the side of journal publishers. Journals too wish to make an impact, and most seek articles that make interesting, positive claims rather than so-called "negative" results. In other words, articles that show some interesting or important correlation are preferred to those that show that no such correlation exists. Both, however, are valuable to science. Publication bias means that researchers may either unconsciously or consciously seek correlations where there are none, knowing that their chances of publication improve significantly with some stated correlation.

Alan Sokal, who was one of the authors of the study above debunking the "happiness ratio," was involved in a noted fraud in 1996. In that case, he was curious about the peer review and publication phenomenon that appeared to result in a number of humanities journals taking matters of hard science as though the domains of the sciences were irrelevant to the humanities, or even harmful. He submitted an article entitled Transgressing the Boundaries: Towards a Transformative Hermeneutics of Quantum Gravity" which argued that quantum gravity, an emerging theory in physics, had progressive political implications. The article essentially argued that traditional science and its methods were part of a cultural hegemony and should be rejected in favor of a "liberatory" science, free from the dogma of the Enlightenment and its cultural biases. The article was complete nonsense, and Sokal wrote and sought to publish it to see if a journal in the humanities would see it for what it was, or choose to publish based upon ideological notions regardless of scientific theory. It was accepted and published in *Social Text*, a journal specializing in post-modern social theory. Shortly thereafter, Sokal revealed his fraud and caused a bit of an uproar in academic circles.

Some have since criticized the "Sokal Affair" for a lack of ethics, namely in the deception of the journal editors via the submission of a fraudulent article. But for the purposes of this discussion, the Sokal Affair illustrates publication bias. The article proposed a positive (though ridiculous) set of claims, and Sokal was a well-known author. Both facts may have contributed to the article's publication in ways that ought not to occur if the ethos of science is properly upheld. There was a lack of proper "disinterest" and thus equipoise, leading to decisions that were not suggested by the actual "research." This phenomenon may help explain why worse frauds are committed and how for some it can even become a modus operandi.

2.5 Fraud

Sometimes the manner and intention behind manipulation of data goes beyond carelessness, negligence, or recklessness. Sometimes it is outright fraud. No modern case more clearly highlights this than does that of Diederik Stapel. Stapel is known for having committed one of the biggest strings of scientific frauds, with the most publicity surrounding his case of any in modern times. At the top of his career as a behavioral psychologist he was a full professor with affiliations to three major research universities in The Netherlands. Among his numerous later retractions was an article in *Science*, one of the two leading, highest impact journals there are. Along with dozens of his papers, the *Science* article relied on fraudulent data. The retractions of his articles and chapters did significant damage to science, including to his students, his co-authors, collaborators, funding agencies, and others in the scientific community who had relied on his sometimes apparently ground-breaking work. In many cases, the data was almost entirely made up, fabricated out of thin air. We know perhaps only because some of his former students or collaborators "blew the whistle" on him. Because he kept close guard over his data, we might never have known unless and until, like Millikan, his notebooks or databases were discovered at some point in the future.

Stapel describes his move toward fraud as motivated by a number of factors. He was deeply interested in social psychology, but frustrated by the messiness of the data it collected and revealed, and with trying to fit that data into coherent theories. Moreover, he was driven by ambition, wanted to be great in his field, to be recognized and quoted and cited and adored. So, he began at first by manipulating data in spreadsheets, a few numbers here and there, to better fit what he wished to convey, similar perhaps to Millikan except that instead of throwing out data that didn't quite fit the hypothesis, he changed it. Eventually, he just started making up data completely to suit hypotheses that he assumed were true, and continued publishing in high-value publications, and remained noted and admired at the top of his field, continuing to bring in research grants, and lecturing and mentoring students at prominent Dutch universities. As of now, he has retracted 54 papers. Among his research was a paper that alleged that meat eating humans were more selfish than vegetarians, which appears to have been based entirely on fabricated data.

The damage caused by Stapel's fraud is significant and continues to hamper the reputation of the entire field of social psychology. Moreover, since his papers have been cited hundreds of times by others, the science upon which those who cited him in support of their own work is undermined. Millions of Euros were spent on his research, and then tens of thousands at least in the various inquiries, and although Stapel deflected criminal prosecution by agreeing to community service and forfeiting some of his pension, the millions he took in grant money over the years to support what was apparently largely fraudulent work did not go to more deserving, possibly fruitful science and will never be paid back.

2.6 Conclusions

It is clear that the ethos of science demands careful relationships between scientists, their data, and the community of researchers. Because science is an inherently communal activity, and because it demands that we remain in a state of equipoise, we cannot work under the assumption that our hypotheses will prove true, and must carefully record, acknowledge, consider, and then reveal the data that is gathered and explain cogently how it leads to our conclusions. Failing to recognize these duties violate at least two of the four ethos of science noted above.

Obviously there are a number of institutional pressures that impel the sorts of misconduct discussed above, including pressure on our careers, the nature of publication and some of the institutional problems discussed already (like publication bias) that distort the publishing business. However, the institutions of science become injured, and the relations between science and the public perverted or fractured every time such misconduct is revealed. This is a threat to science itself, and also to the public with which it has a mutually-beneficial, though dependent relationship. Scientists should keep the above cases in mind, consider how they effected not just the scientists involved, but the institutions, other stakeholders like the public, and the community of scientists and colleagues who are all supposed to be working toward one goal and one goal only: the truth.

Study and Discussion Questions

1. How does the ethos of science demand proper citation and attribution of authorship? What happens to science and other scientists in the case of failure?
2. Describe some criteria that are appropriate for considering someone an author. Name some that are inappropriate.
3. What is publication bias? How does it improperly affect science? Describe the ethical errors involved in the "happiness" ratio study.
4. What social and institutional pressures might be encouraging fraud, and how can we help to avoid or overcome them?

References

Ashman, Keith, and Phillip Barringer. 2005. *After the science wars: science and the study of science*. London: Routledge.

Brown, Nicholas JL, Alan D. Sokal, and Harris L. Friedman. 2013. The complex dynamics of wishful thinking: The critical positivity ratio. *The American Psychologist* 68(9): 801–813.

———. 2014. The persistence of wishful thinking. *The American Psychologist* 69(6): 629.

Callaway, Ewen. 2011. Report finds massive fraud at Dutch universities. *Nature News* 479(7371): 15–15.

Claxton, Larry D. 2005a. Scientific authorship: Part 1. A window into scientific fraud? *Mutation Research/Reviews in Mutation Research* 589(1): 17–30.

References

———. 2005b. Scientific authorship: Part 2. History, recurring issues, practices, and guidelines. *Mutation Research/Reviews in Mutation Research* 589(1): 31–45.

Escalante-Ferrera, Ana Esther, Luz Marina Ibarra Uribea, and César Darío Fonseca Bautistab. 2015. "Questionable" Behaviors and Practices in Academic Productivity in Postgraduate Studies in Mexico. *Sociology* 5(1): 8–22.

Evans, James. 1987. On the origin of the Ptolemaic star catalogue: Part 1. *Journal for the History of Astronomy* 18(3): 155–172.

Gingerich, Owen. 1980. Was Ptolemy a fraud. *Quarterly Journal of the Royal Astronomical Society* 21: 253.

Gross, Paul R., and Norman Levitt. 1997. *Higher superstition: The academic left and its quarrels with science*. Baltimore: JHU Press.

Guillory, John. 2002. The Sokal affair and the history of criticism. *Critical Inquiry* 28(2): 470–508.

Hilgartner, Stephen. 1997. The Sokal affair in context. *Science, Technology & Human Values* 22(4): 506–522.

Holton, Gerald. 1978. Subelectrons, presuppositions, and the Millikan-Ehrenhaft dispute. *Historical Studies in the Physical Sciences*, 161–224.

Jennings, Richard C. 2004. Data selection and responsible conduct: Was millikan a fraud? *Science and engineering ethics* 10(4): 639–653.

Levelt, Willem J.M, P.J.D. Drenth, and E. Noort. 2012a. *Flawed science: The fraudulent research practices of social psychologist Diederik Stapel*.

———. 2012b. *Flawed science: The fraudulent research practices of social psychologist Diederik Stapel*.

Macrina, Francis L. 1995. *Scientific integrity: An introductory text with cases*. Washington, DC: ASM Press.

Niaz, Mansoor. 2000. The oil drop experiment: A rational reconstruction of the Millikan-Ehrenhaft controversy and its implications for chemistry textbooks. *Journal of Research in Science Teaching* 37(5): 480–508.

Rennie, Drummond, Veronica Yank, and Linda Emanuel. 1997. When authorship fails: A proposal to make contributors accountable. *Jama* 278(7): 579–585.

Stroebe, Wolfgang, Tom Postmes, and Russell Spears. 2012. Scientific misconduct and the myth of self-correction in science. *Perspectives on Psychological Science* 7(6): 670–688.

Chapter 3
Issues of Authorship

Abstract Publishing is the currency of academia, it is in large part the measure of the worth of an investigator in any academic field. The need to publish, combined with other institutional pressures for funding, promotion, etc., may contribute to temptations to be named as authors inappropriately or worse. Institutional norms, and diverging norms among various fields, make the landscape for rules about authorship complex. To what degree and for what reasons must authors be named and in what order? What scientific norms demand which forms of behavior by an author regarding truth? What counts as authorship, and why is this important to science? I explore these issues below and offer some guidance for authors concerned about conflicts with norms of authorship within and among institutions, taking cues from the Mertonian norms discussed above.

3.1 Publish or Perish

Scientific publication is the fundamental unit of value in the various professions of science, and the essential means of ensuring that the methods of science work over time. In other words, the manner by which hypotheses become tested, challenged, confirmed, or falsified over time is through the medium of written words, and authorship is a complicated matter both for ethical and practical reasons. Where the words come from, whose words they are, and most importantly, who takes responsibility for the words and what they represent is an enormous issue in science, and offers numerous opportunities for error, harm, and thus ethical reflection.

A scientific work of authorship is the cumulative result of a research program's reaching some conclusion sufficient enough to warrant dissemination to the community of researchers. When results are published, it is in respect of the ethos of science, particularly the values of communalism and organized skepticism. The scientific purpose is to expose results to testing, to afford other researchers an opportunity to challenge results so that they can either confirm or falsify some hypothesis or theory. While scientific publishing has accumulated a number of institutional roles beyond the fundamental duties of science (as discussed, the search for truth), its primary role should remain the slow and deliberate investigation and description of natural law. The authors are generally those who have participated most closely with the experiments discussed, and who have in some way contributed to advancing progress toward knowledge of the laws of nature. There are a number of duties that come with those roles, and in publishing a number of new duties arise. These duties multiply as the number and types of stakeholders change. A scientific paper involves not just the author, but also co-authors, fellow researchers, funding agencies, host institutions, as well as the scientific community and the public at large. Given the web of stakeholders, it is natural that the various duties have become increasingly complicated, and unsurprising that a number of public embarrassments to science, as well as worse harms, have flowed from failures to properly consider and abide by the duties of authorship.

3.2 Duties to the Truth

The primary duty of authorship in science is to the truth. As discussed previously, science works in part by accepting a form of realism in which we assume that underlying nature are knowable, consistent, universal laws and that the object of science is improved understanding of the working of nature. There is some truth, apart from our own interpretation, and it can be known and described in ways that others can test. Of course, the truth is often obscure and nature's laws difficult to comprehend much less observe consistently. As with the experiments conducted by Ehrenhaft and Millikan to discern the charge of the electron, sometimes the state of the art means that the reach of a scientist exceeds her grasp, and we catch only glimpses or some vague notion of the truth. When that is the case, the duty to describe what one finds is especially pronounced. Authors must be transparent. They must present their findings in ways that are intelligible and that admit of replication.

It is clear in the case of Millikan, and with the benefit of hindsight and further evidence, that his publication did not adhere closely enough to the truth. He failed in his duties to disclose, and in fact outright lied in the text. Those who would try to replicate based upon his paper were erroneously led to the false conclusion that their own experimental results were hopelessly flawed, and thus potentially valuable scientific data was often discarded. Millikan was the sole author of his article in 1913 in the *Physical Review*, and sole proprietor and reporter of his data, although he acknowledged the assistance of a Mr. Lee in making some of the observations. What we know is that although Millikan claimed that he reported all of the runs of the experiment, he did not. He only reported about half, presumably the ones he felt were most representative of the apparatus working well, and those which so happen to most closely conform to his expectations. Millikan, by taking full and sole authorship of the paper, takes also all of the responsibility, and also the blame for his ethical failures. The harm he did to science, as we have noted, is mentioned by Feynman and recognized by others. Instead of aiding progress toward the truth, he sprinted to a conclusion that turned out to be true, or very nearly true, without allowing for the proper testing and challenge of others due to his not disclosing the full truth.

The truth demands of the scientific author that an experiment can be replicated, that the data reported is the data observed, that any failures of shortcomings are properly noted, and that to the best of the scientist's ability, he or she is taking sufficient account of all of the factors that are relevant so that others may challenge, test, confirm, or falsify without wasting inordinate time or effort. As well, the truth demands transparency not just of data but of language. Scientific papers should be easily understood by others in the field, and obscure and difficult to penetrate language cannot prevent experts from comprehending the nature of the study, nor replicating and challenging its results. Communicating the truth understandably is as important as revealing the methods and data that are alleged to underlie it. Duties to the truth imply duties to the scientific community and to the public as well as other stakeholders. The thread of duties is easier to trace when there is a sole author, but in most science conducted today, scientific papers are the result of elaborate collaborations and numerous parties, often geographically dispersed. Let's consider next the various duties that may arise as the number, nature, and even location of authors increases, sometimes significantly.

3.3 Duties of Authors to Each Other

Most modern science takes place across research groups involving numerous people, sometimes of differing scientific specialties, and often geographically dispersed – not even working in the same physical laboratory. This presents complications to the duties they owe, not necessarily toward the "truth" but often complicated by social, cultural, and discipline-specific practices regarding authorship and acknowledgement.

When there is one author, the question of who counts as an author is elementary. However, when more than one author is involved, the question becomes more difficult. Disciplinary practices and conventions, as well as institutional ones, further complicate the matter. Authorship's various institutional roles include: status, prestige, honors, career advancement opportunity or indicator, and others. Being named as a primary author, or in some cases described as the "first author" is for a large number of fields an important qualification or honorific. In the current scientific milieu, with the development of the h-index and other similar measures of academic merit or worth, authorship plays a significant role. There are thus many pressures on scientists not only to be named as authors of articles, but to have specific placements in the order of authorship. Taking on the role of an author among other authors is itself assuming a position of responsibility, both to the truth, and to the other named authors. The first or corresponding author typically takes on the bulk of the responsibility, but others who take on the benefits of being named authors cannot thus shirk their own responsibilities for the contents of a paper.

Authors of scientific papers often have differing responsibilities based upon their particular areas of expertise. Complex research programs now frequently rely upon non-overlapping specialties, and no one scientist may be expected to have sufficient expertise in all areas to complete a study. Collaborating scientists must then rely upon the expertise and honesty of their collaborators, and their representations of their particular areas of the study. It is thus incumbent upon collaborators to be fully transparent, open, and forthcoming primarily with one another in the course of writing and publishing, and to ensure as best they can that their collaborators have at least a basic grasp on their contributions. Failing this may mean failing the collaboration. As we will see below, in case there is something terribly wrong with the paper itself, no collaborator can avoid the responsibility for error, fault, or fraud just because the error or fraud originated with a collaborating author. Taking on the benefit of being an author also means taking on the risks involved in case of co-authorship. All too often, papers are retracted due to fraud or gross error, and being a co-author of a retracted paper reflects negatively on all of the co-authors. It will not suffice to claim that one did not know of the fraud or recklessness of a co-author, since in the course of a collaboration all of the parties will have benefitted from the institutional value of their co-authorship. This is so regardless of the level and quality of the collaboration, and so we should examine what in the first place qualifies someone as an author. What level of contribution is necessary in order to name someone as an author in the first place?

3.4 Are You an Author?

Discovering authorship is easier without co-authorship, and in the case of co-authorship may become increasingly difficult with an increasing number of authors. Whether you should be counted as an author for a scientific publication, as mentioned, often depends upon some relatively disciplinary-specific conventions, and it is not unusual to find named authors who have not necessarily even read the whole paper, much less contributed to its writing. Department chairs, laboratory heads, and others have become accustomed to being named as co-authors by virtue of those positions and their authority, with little resistance by their fields or institutions where this sort of practice may have become accepted and perceived as correct practice. But there are ethical hazards associated with authorship, and one way to approach the question is to ask: who takes responsibility for the paper, chapter, or book? Many people are willing to take responsibility for work which, although they may not have amply contributed to it, leads to praise, but the important question is: who will take responsibility in case a work leads to criticism, or worse?

Not everyone who counts as an author need have put pen to paper (or tapped a keyboard), but there are some minimal acts that must count toward authorship, without which certain disciplinary practices involving naming authors based on authority must be questioned. A paper is more than the collection of words between its

3.4 Are You an Author?

beginning and its final period, it is an expression of ideas, and in science the meaning of the paper (or book, or chapter) is the result of some set of observations and what they imply for some hypothesis. In the context of a research program, the meaning of the scientific work is either some confirmation or falsification of some hypothesis, and the strengthening or demolishing of some theory. One can contribute to the scientific role of a paper without having written a word, simply by proposing a significant portion of the ideas that generate the experiment, or describe its results, or put it in a larger context of a discipline, its theories and axioms, and the future for study based upon the described study. The "author" may thus have contributed mostly to the intellectual basis, theoretical foundation, and creative or exploratory impetus behind a study, providing the direction of the research team, leading discussion and analysis of the data, and coalescing the results into a scientifically relevant product. But the "idea" person, as named author, must still do one more critical thing: read and comment on the paper, even if he or she never wrote a word of it.

More typically, the first author is also the major contributor to the actual verbiage in the paper. This is preferable. The first author (and we'll talk a bit more below about the order of authors) is also typically the responsible, or "corresponding" author, who will be the conduit between editorial discussions at the journal or press, and the authors involved. Having spent the most time informing the ideas, the direction, the actual verbiage of the paper ought to count significantly toward "first" authorship, but any combination of these sometimes suffices according to various institutional or disciplinary standards. It is not unheard of that a first (or last) author has done a fair amount of the intellectual chore of devising the research, assigning tasks, and editing the verbiage of the paper, without necessarily "writing" the paper per se. Whether this ought to be the case is another story and will be discussed below. It is perhaps helpful to first discuss what is NOT sufficient for authorship, and why.

A position of authority is not enough. Simply being in charge of those who write a paper is insufficient to warrant the claim of authorship, despite institutional or disciplinary practices. Something more, such as some intellectual contribution to a particular work must be present. Again, we will examine below some of the hazards that come from ignoring this advice. Simply reading a paper is also insufficient. Although every named author ought to at least read the paper they are named on, it is not sufficient for the claim of authorship. Proofreaders or even editors are not authors. Authors create something, and while proofreaders and editors contribute to the shape of the creation, they are not necessarily the creators of the final product. Funding is not enough. Although those who procure grants that enable the research behind a paper are often counted as last authors, the mere procurement of the grant is insufficient on its own. Typically, those who get the money have also contributed to the ideas behind the paper, some major hypothesis, the design and conduct of the study, etc. These, combined with the procurement of the grant may suffice, and often (depending very much on disciplinary and institutional practices) get to be named as final author.

Providing lab space or equipment is also not enough. Again, although this happens far too often, the mere allocation of resources does not equal authorship. Helping or gathering the data using that equipment may well suffice for authorship, as it contributes to the testing of some hypothesis, especially if combined with analysis. Too often, those who provide lab space or other resources become named authors as a means of recognizing their generosity, but it is in the breach that the dangers of this sort of practice are revealed. When something goes wrong, how are we to allocate blame? If credit is too readily handed out, it should be understood that the blame for failures will be similarly allocated, and in all likelihood, every named author will suffer. Let's look at a prominent example, the case of Hwang Woo-Suk.

3.5 The Korean Stem Cell Fraud

Dr. Hwang Woo-Suk was a famous and well-funded researcher in the area of cloning and stem cell research, based at Seoul University in South Korea. In 2005, an apparently ground-breaking paper authored by him and 24 co-authors was published in *Science*, and claimed to describe the successful cloning of eleven embryonic stem cell lines from a single subject. The breakthrough guaranteed Hwang Woo-Suk national heroic status in Korea, helped him secure yet more funding and other laurels, and made South Korea a center for the burgeoning field of "therapeutic" cloning. A number of ethical problems then came to light about the research and then the paper, including that the donor cells were taken from students who were paid for their contributions, specifically from their eggs. This may have been a conflict of interest and abuse of authority in the lab, and at least one of Hwang Woo-Suk's co-authors claims to have been misled about the source of the donor eggs. That co-author, Gerald Shatten, was subsequently investigated by his institution, the University of Pittsburgh, and found guilty of misbehavior (not misconduct) in failing to properly conduct his responsibilities as a co-author, including failing to catch certain inconsistencies, and failing to ensure that all other co-authors has approved of the paper before publication. Things became much worse for all involved a few months later when *Science* received an anonymous tip that two of the photos used in the *Science* article were in fact duplicates, when they were represented as having been of two different stem cells. After this, one of the co-authors, Sung Roh, confessed to the media that Hwang Woo-Suk had admitted privately that data regarding nine of the eleven cells lines alleged to have been cloned were fabricated. Further inquiry revealed that all of the data had been fabricated, and that improper authorship was given for the mere act of contributing oocytes. Even though Hwang was credited with the world's first cloned dog, following the stem cell scandal, he was forced to resign his position and faced criminal prosecution for fraud and given a two year jail sentence for related misconduct.

This case reverberated throughout the international research community, revealing one of the major, most publicized and expensive scientific frauds of modern

3.5 The Korean Stem Cell Fraud

times, and for the purposes of our discussion in this chapter (though indeed we could discuss any number of issues related to scientific integrity throughout this book, focused on this case alone), significant issues relating to authorship emerged. As mentioned, one of the co-authors was specifically punished for his role, but there were clearly a number of negative repercussions for all the co-authors, not least of which was to be associated with such a blatant and disreputable case of fraud. Most prominent journals require co-authors to sign statements about their agreement with the conclusions of the paper they co-authored, as well as to provide descriptions of their contributions. Further changes were proposed in light of the Korean case, especially given the apparent payment via co-authorship for the procurement of tissues. A new proposed policy of the International Committee of Medical Journal Editors, for instance, would suggest a number of necessary factors for co-authorship, including taking personal responsibility for a paper's contents and conclusions. The proposed policy spells out other specific conditions, including:

1. substantial contribution to conception and design of a study, acquisition of data, or analysis and interpretation of data
2. drafting the article or revising it critically for important intellectual content
3. final approval of the version to be published (these first three are necessary)
4. all authors must further complete and submit an authorship form with signed statements on responsibility, contributions, financial disclosure, and copyright transfer or federal status

By stressing the notion of personal responsibility, the hope is that authors who are willing to take the credit are also aware of, and willing to accept their roles as responsible parties. In the Korean stem cell case, as is probably the case with too many studies, authorship was bartered for some benefit, not allotted according to scientific criteria, and ultimately harmed all involved.

Although Hwang was primarily responsible for the stem-cell fraud, his co-authors share responsibility by virtue of being co-authors. This is unavoidable both ethically and socially. It is no defense to claim that one has taken the benefit of co-authorship without also taking on the responsibility it implies. The implication is the result of institutional norms (such as signing off on the contribution) and by the ethics of authorship. Authors take the role of creator, accept the duties of creating, and the blame if something is wrong with their creation by virtue of their negligence or intention. The reward of authorship was dangled like payment toward those who ill-considered the effect that taking that honor carried. Authorship is treated like a reward in part because of the changing nature of academia, in which one's publications and resulting "h-index" are used as measures of one's academic work, and sometimes salary and promotion choices hinge upon that measure. Authorship is not an honor. Although it reflects due to various institutional decisions and arrangements upon our perceived value as scientists, it I primarily an acknowledgment of responsibility. The author is answerable for the study, the methods, the conclusions, and the expression of all of these to an audience, and the purposes of science are the primary motivation for ethical publication.

3.6 What Counts as Your Work?

There has been increasing focus, especially in the humanities but also in the sciences, on the problem of plagiarism. As with failing to provide appropriate sources so that future researchers can trace back one's work, using the words of another poses problems for scientific integrity. Technically, whenever one uses six or more words in a row written by another person, one has a scientific and moral duty to attribute the source. In our modern age of digital publications with the ready ability to copy and paste text among documents, it is especially prevalent. In the most egregious instances, "authors" have submitted entire works copied from another, replacing only their names and passing off the work as their own. Lesser forms of plagiarism still abound, despite a number of readily-available tools online for checking whether a paper has been plagiarized.

Because scientific publishing is in part a representation and record of a scientist's contributions to a field, there is both the scientific necessity to properly attribute and the moral duty not to pretend to have done work one has not done. Research, once published, is a living, active entity in a discipline. At its best, it becomes the source for the work of others to build on, or to falsify and move on, and so the connection between its originator must not be severed. The expression of scientific hypotheses and results too must be traceable to the author as the wording matters. The ways we express our findings may have ambiguity, may require clarification, may be subject too to challenge, as the author of a paper remains responsible for her words once published. When one uses the words of others, the ability to answer for the means and manner of their use is severed from the true author. Moreover, passing off another's expressions for one's own, without attribution, deceives the audience into thinking you have originated a particular expression. It denies the true, first author the moral right to be identified as the author.

Consider the case of David Davies' book *The Last of the Tasmanians* published in 1973. This ethnography of the now extinct Tasmanian natives gives details, includes interviews, and provides compelling background on a race of people who all but disappeared from their native home, lost their native tongues, and whose last 47 members were transferred from Tasmania in 1847. Davies' 1973 account is a thorough account, useful for ethnographers, and filled with details that have thankfully been preserved for history, just not by Davies. Although he cites a work by James Bonwick, Davies fails to cite his essential work of Tasmanian ethnography: *The Lost Tasmanian Race* (1884). A comparison of the Davies book and Bonwicks, done by Dr. Charles F. Urbanowicz in 1998 and published online (http://www.csuchico.edu/~curbanowicz/Pacific/Tasmania.html) puts significant portions of the book side by side for comparison. Here is but one example from Urbanowicz's paper:

3.6 What Counts as Your Work?

BONWICK 1884, page 3	DAVIES 1973/4: p. 13
"The discoverer of the island, Abel Jansen Tasman, never saw the original inhabitants. He detected notches in trees by which they ascended after birds' nests, as he supposed, after opossums, as we know. He did observe smoke, and heard the noise of a trumpet. Satisfied with hoisting the Dutch flag, he passed on to the discovery of New Zealand. A Frenchman, Captain Marion, held the first intercourse with the wild men of the woods. This was in 1772, being 140 years after Tasman's call. Rienzi, the historian, speaks of the kind reception of his countrymen by the Natives, whose children and women were present to greet the strangers. But bloodshed followed the greeting. This is the account:--'About an hour after the French landed, Captain Marion landed.'"	"The discoverer of Van Diemen's Land, Abel Jansen Tasman, never saw the original inhabitants. However, he detected the notches put in tree trunks by which they climbed the trees looking for, he thought, birds' nests, though later evidence showed that it was for opossums. He often saw smoke, and heard a noise like a trumpet (the blowing of the conch shell). Tasman was satisfied with hoisting the Dutch flag, and then he sailed on to discover New Zealand, much more welcome to an explorer. A Frenchman, Captain Marion, made the first contact with the wild men of Van Diemen's Land, in 1772, 140 years after Tasman's landing. Rienzi, the historian speaks of the kind reception the natives gave his countrymen. Women and children were present to greet the strangers, which indicated that they did not have war on their minds. But a little later there was bloodshed: 'About an hour after the French fleet had landed, Captain Marion landed.'"

One can read through both books and have a tremendous feeling of *deja vous*. Instances like this one above abound. The language appears almost completely copied, except for some modernization of language style, and yet the Bonwick work from 1884 is never credited. There is no copyright violation because the copyright on Bonwick's work had already expired, but it appears to be a case of plagiarism of almost an entire book. Here is now the Australian library system currently describes Davies' book (http://trove.nla.gov.au/work/21566816):

The last of the Tasmanians / [by] David Davies
Davies, David Michael, 1929-
View the summary of this work
Author
Davies, David Michael, 1929-
Subjects
Aboriginal Tasmanians - Government relations.; Aboriginal Tasmanians.; Mosquito.

Summary

Plagiarized with slight modernization of style, from J. Bonwicks The Last of the Tasmanians, 1870, which see for complete annotation; Davies has added a new concluding chapter, Origins of the Tasmanians in which he argues the possibility that Tasmanians represented a separate human species and for a multiple cradleland theory in which the various races are believed to have; New Appendix; Extract from Mrs. C. Merediths My Home in Tasmania, 1852.
Bookmark
http://trove.nla.gov.au/work/21566816
Work ID
21566816 [emphasis added]

Clearly, having such a reference for one's book harms one's own academic and scientific standing. But there is a further harm to science from this sort of act. Suppose that one has access to the Davies book but not the Bonwick book (which, it should be mentioned, is rather thoroughly sourced). The facts alleged, based in some cases on interviews with surviving sources (in the 1800s), cannot be traced back. They cannot be challenged, compared, or evaluated for their authenticity or accuracy. The link to the original data has been fatally severed, and science is harmed in its future progress.

3.7 Salami Science and Self-Plagiarism

There is another form of plagiarism that does harm in science: so called, "self-plagiarism". This is the act of taking one's own words from previous publications and republishing them in another work without properly attributing their origin. Although the moral harm of taking credit for another's expressions is not a risk in self-plagiarism, the risks and harms to science are still present. The impetus for this practice is also clearly the institutional expectations and reward associated with the number and impact of publications, as well as the h-index which helps to measure a scientist's impact in a field. The tendency then to split up a study into numerous parts and publish it as more than one paper, called "salami science" is not plagiarism per se, but can often be spotted by it. An author who wishes to publish a study in multiple journals, perhaps by severing it into some logical although too-closely connected parts, may find that sections of one paper are thus easily re-used in the other paper. It is easier to cut and paste sections, for instance, describing methods from one paper to another. Without attributing the language to the other paper, however, the "salami slicing" of a study is harder to detect.

Failing to acknowledge that the original source of words in a paper came from another paper, even one's own, deceives the audience into thinking that the verbiage is as it should be: original, or cited. This too severs the connection between the first instance of investigation and expression and the one being passed off as new and original. The deception is also to the publishers, the journals often run by volunteers, who solicit the promise and trust the author that the words and science being submitted are original. Self-plagiarism and its relative salami science waste precious resources and violate the trust that is necessary for scientific publishing to function. Readers and other scientists too may rightly feel their trust violated by these practices.

3.8 Conclusions

Authorship is a responsibility. The scientist who takes on the mantle of honor and creativity, as well as the associated institutional and cultural rewards afforded by authorship, must also be accountable for their work. Accountability for the sake of science means that other scientists can trust that the authors know as much as possible about their own work, have verified that their words properly represent it, and that all those claiming to be authors are personally responsible for their contributions as well as the work as whole. When the real source of words or data is not properly given, then the author fails the trust and duty owed to the community of scientists, and violates the ethos particularly of communalism, disregarding the necessity of the community of researchers to be able to test and verify, as well as trust the veracity of what they read.

Because questions of the order of authorship vary among disciplines, one must work these arrangements out well beforehand in order not to have disagreements later among the authors. Authorship cannot be traded or bartered, but must depend upon actual contributions. One means of testing whether someone should properly be considered an author is to ask whether the paper as it exists would exist as it does without their contribution. Morally, each of those accepting the status of author should also ask themselves whether, although they are willing to accept the benefits of authorship, they are also willing to accept all the risks in case ofsome error, or worse, is present in the work.

Study and Discussion Questions

1. What duties does an author owe and to whom?
2. What must the absolute minimum activity for authorship be? How should co-authors determine ordering of authorship and whether one should be an author on a particular paper?
3. How might co-authors resolve issues of authorship best and at what stage of the authoring process?
4. How do ethical lapses regarding naming and claiming authorship relate to the Mertonian norms?
5. What is the ethical problem with "self-plagiarism" and how best to correct it?

References

Bennett, Dianne M., and David McD Taylor. 2003 Unethical practices in authorship of scientific papers. *Emergency Medicine* 15:(3): 263–270.
Bonwick, J. 1884. Tie Last of the Tasmanians. 1870. Daily Life and Origin of the Tasmanians. 1870. *The Lost Tasmanian Race.*
Cho, Mildred K., Glenn McGee, and David Magnus. 2006. Lessons of the stem cell scandal. *Science* 311(5761): 614–615.
Coats, Andrew J.S. 2009. Ethical authorship and publishing. *International journal of cardiology* 131(2): 149–150.
Davies, David Michael. 1973. *The last of the Tasmanians.* L. Muller.
Drenth, Joost P.H. 1998. Multiple authorship: The contribution of senior authors. *JAMA* 280(3): 219–221.
Faunce, Thomas Alured. 2007. Whistleblowing and scientific misconduct: Renewing legal and virtue ethics foundations. *Journal of Medicine and Law* 26(3): 567–584.
Holton, Gerald. 1978. Subelectrons, presuppositions, and the Millikan-Ehrenhaft dispute. *Historical Studies in the Physical Sciences*: 161–224.

Kitzinger, Jenny. 2008. Questioning hype, rescuing hope? The Hwang stem cell scandal and the reassertion of hopeful horizons. *Science as Culture* 17(4): 417–434.

Koepsell, David. 2012. Some ethical considerations in astronomy research and practice. *Organizations, People and Strategies in Astronomy Vol. 1, Edited by Andre Heck, Venngeist, Duttlenheim (2012)* 1: 265–274.

Laflin, Molly T., Elbert D. Glover, and Robert J. McDermott. 2005. Publication ethics: an examination of authorship practices. *American Journal of Health Behavior* 29(6): 579–587.

Marušić, Ana, Lana Bošnjak, and Ana Jerončić. 2011. A systematic review of research on the meaning, ethics and practices of authorship across scholarly disciplines. *Plos one* 6(9): e23477.

Osborne, Jason W., and Abigail Holland. 2009. What is authorship, and what should it be? A survey of prominent guidelines for determining authorship in scientific publications. *Practical Assessment, Research & Evaluation* 14(15): 1–19.

Resnik, David B., Adil E. Shamoo, and Sheldon Krimsky. 2006. Commentary: Fraudulent Human Embryonic Stem Cell Research in South Korea: Lessons Learned. *Accountability in Research* 13(1): 101–109.

Rogers, Lee F. 1999. Salami slicing, shotgunning, and the ethics of authorship. *AJR. American Journal of Roentgenology* 173(2): 265–265.

Sandler, Jeffrey C., and Brenda L. Russell. 2005. Faculty-student collaborations: Ethics and satisfaction in authorship credit. *Ethics & Behavior* 15(1): 65–80.

Urbanowicz, Charles F. COMMENTS ON TASMANIAN PUBLICATIONS OF 1884 AND 1973/74. http://www.csuchico.edu/~curbanowicz/Pacific/Tasmania.html

Chapter 4
Issues in Intellectual Property and Science

Abstract Copyright, patents, and trademarks are the most common forms of "intellectual property" we confront. Scientists, researchers, and academics must often be aware of and careful about each of these at some point in their careers and while pursuing their research. Who owns what, or has what sorts of control over various expressions and objects, may influence one's abilities to publish about, or research phenomena. Navigating those rules of ownership requires at least a basic understanding of what counts as intellectual property, the ideas of "fair use" that enable one to use otherwise protected matters in certain ways, and working with experts to be sure that one's work and their institutions interests are properly considered and accommodated. Here I will introduce those concepts, provide some examples, and lay out some potential conflicts and issues of concern for researchers.

4.1 What Is Intellectual Property (IP)?

Intellectual property is a class of legal protections first created about 250 years ago. Before the laws of copyright and patent were invented, there were no means for creators of either aesthetic or utilitarian works to prevent others from creating the same things. Shakespeare, for instance, never asked the state's help to prevent other from performing his plays, though he did attempt to prevent others from doing so by other means. There was no law that prevented anyone who wished from putting on a competing production of Hamlet. Archimedes could not prevent others from using his ideas about moving water to use, nor could he stop them if they created the same mechanisms completely independently. This is because, without some law to protect a creator, ideas can be freely exchanged and used without depriving anyone of their property. Prior to intellectual property (IP), property laws protected things that are naturally excludable. Things that counted as property, as ownable, were only those things whose possession naturally excluded the simultaneous possession by another. "Real" property, which includes moveable objects and land, is naturally excludable because when one person (or sometimes a group by collective effort) possesses or occupies it, in order for another to take possession some other act must intervene to dis-possess the prior possessor of their possession. As an example, if I am driving my car and another person wishes to possess it, they must kick me out of it, lease, buy, or borrow it from me. The objects of IP don't work like this.

The ideas and the story, as well as the particular verbiage of Hamlet can be held, said, and written by any number of people without infringing on Shakespeare's possessions. Unlike with real property, the simultaneous usage of an idea by more than one person does not deprive anyone of any usage rights or abilities. This is in fact part of the value of creativity, its ability to replicate and spread and do so without diminishing, but rather enhancing the shared wealth of all, increasing therefor the intellectual commons. Such openness is part of the value and mechanism of science as well. Testing, replicating, and improving upon the works of others is essential to the ethos of science, which is communal and universal. Intellectual property is part of our modern economic landscape, and has become essential to technological development and creativity in aesthetic media, and so we should be wary of how it interacts, and may present ethical issues for, science and its institutions. Let us look at IP laws, their evolution and modern forms, and then see how they impact ethics in science.

4.2 A Short and Sweet History of IP

As we discuss above, property used to be a concept reserved for "real" (*res*, latin.) property which was limited to objects and areas that are somehow excludable. But social institutions and norms have improved on and altered the ways we delineate our "ownership" of the things we possess, for instance creating licenses, easements, sovereign rights, eminent domain, joint and several ownership, as well as other permutations that expand upon mere possessory rights. Meanwhile, inventors and creators of aesthetic works have devised numerous ways to improve society and states began to be interested in ways to both attract and reward their improvements. Early in Renaissance Italy, Venice created an incentive for inventors to immigrate to Venice, devising a monopoly right for them in their inventions. It was the first "patent," to be imitated eventually in other European countries. "Letters Patent" came to be favored in England in the early Enlightenment era, and bestowed by grant of the sovereign (king or queen) an exclusive right for an inventor to practice their art (some invention, presumably, or process) within the bounds of the sovereign's domain. This monopoly right given by the crown meant that no one could compete with the patent holder during the time of the patent. Having the state clear all competition from one's path is a serious incentive (it has long been believed) for innovation and creation, and rewards the state by encouraging innovation, while rewarding the creator with the exclusive rights to the invention or expression.

Letters patent were eventually curtailed by the British Parliament, with the passage of the Statute of Monopolies, given that the crown was perceived of as abusing the system by granting permanent monopolies as favors, reducing competition in the marketplace by doing so, and actually stifling inventiveness. The bludgeon of a monopoly granted by the state with the power of the courts behind it is a weapon that needed to be stifled, and so the Parliament reined in the terms of patents (which had been permanent) and limited them to a number of years (typically 14). By the time the colonies in the Americas gained their independence, and enshrined IP monopolies in powers granted under their new constitution allowing the congress to create patents and copyrights, the favored term of both aesthetic and inventive monopolies was 14 years.

The types of IP that currently exist include: patents, copyrights, and trademarks. Another form of protection that has existed for a long time, pre-dating IP la and still widely employed is: "trade secrets." Anytime someone creates something (a process or some product) that they think they can employ to some competitive advantage without revealing it, they can choose to protect it simply by keeping it a secret. Trade secrets have existed as long as people have been keeping secrets. If you create something that is difficult to "reverse engineer," so that another who uses it will have difficulty replicating it, or you make things yourself using processes that are difficult for others to figure out, then keeping a "trade secret" is a low-cost alternative to state-created monopolies. Trade secrets become harder to keep over time, as generations to come need to learn and replicate the secret art, and as reverse engineering practices and products becomes easier. One way to combat the spread of trade secrets is the evolution of guilds, and in fact guilds employed various means, including intimidation and force, to keep others who threatened to compete in a market with something being kept as a trade secret from posing a threat to their captive market. You either joined the guild, paying some fee and protection money for belonging, or you were forced to go elsewhere outside of the area of influence of the guild.

As science was developing as an institution, at the dawn of the Enlightenment, and scientific journals and salons were evolving to manage the proper dissemination, testing, and spread of scientific ideas, patents and copyrights were emerging as civilized alternatives to guilds, secret-keeping, threats, and force. By taking the protection of an aesthetic or useful art out of the hands of gangs, essentially, and putting enforcement into the courts, as well as limiting the terms of protection, it was thought that inventiveness and creativity would be encouraged. Copyrights protect "non-utilitarian" expressions, including now any written work of fiction or nonfiction that is more than simply instructions, films, audio works, plays, written choreography, and any other such expression once it is reduced to some "fixed medium." Patents cover "inventive" works, including processes and products if they are "nonobvious," "useful," and "new."

Patents and copyright are state-created rights, and their terms of protection have varied over time. As well, the extent of the protection afforded by national laws has changed over time and continues to be adjusted to maximize their intended benefits, as well as according to various shifting political priorities. Although the jurisdiction of a copyright or patents is generally limited to the nation in which it was issued, various international agreements have standardized protections around the world, and created international means of enforcement of the various monopolies granted.

A copyright is typically created by the mere creation of some work in some fixed medium. In other words, as soon as one writes something new, or films or records a new expression, the creator owns the copyright. Usually, one need not file any formal document to record the copyright, though without doing so it could become harder to prove the first creation of the expression if there is some later dispute. Copyrights protect the particular expression of an idea, not the idea itself, as long as there is no "substantial similarity" between two expressions of the same idea, then there is likely no "infringement" of copyright. In case of independent authorship, where two or more people independently author substantially similar works, as long as the authorship was not influenced by the other work (and is considered, innocent rather than copying) there is also no infringement. Copyrights last generally the lifetime of the author plus an additional period of time, generally 70 additional years. Patents work quite differently.

Unlike copyrights, which come into being the moment a work is created, a patent must be applied for with some bureaucracy, and a review process determines whether it will issue. Patent offices generally will only grant patents for inventions that are new, non-obvious, and useful. Valid inventions can be either processes or products. So, for instance, if one creates a new drug, a chemical compound that does not exist naturally, and that is useful, and not just an obvious minor alteration of a known drug or compound, one could presumably get both a patent on the product (the chemical) but also on any unique process by which the chemical is created. Once someone obtains a patent, they can prevent others from reproducing the object or process patented, and maintain a monopoly on it for about 20 years. After the end of the patent, anyone can make or use the patented object or process as they please. In order to get a patent, the "art" must be revealed so that others skilled in that art can reproduce it.

Patents are a tradeoff. Part of the rationale behind the patent system is that disclosure of inventions is helpful for the progress of the sciences. As discussed above, one historically, pre-legal method of maintaining the monopoly over some discovery or invention is secret-keeping in the form of trade secrets. However, the ethos of science requires disclosure. Science does not progress if when something new is discovered about nature it is not properly disclosed and tested by others. The ethos of science demands openness. Yet those who undertake scientific research often have some desire for reward other than those typically associated with science. Not every researcher works in academia. In order to maintain an incentive beyond academic rewards for discoveries and inventions, as well as to encourage disclosure, patents provide a monopoly period during which one can recoup the capital investments behind research and development in exchange for disclosure of the underlying art and eventually the lapse of the technology into the public domain.

Patents are now especially important for scientists due to the ever-changing nature of academic funding and collaborations, and the emergence within the past few decades of "technology-transfer" offices at major research universities attests to the commitment by most academic institutions to a future in which patenting technologies developed through basic research is being embraced. Increasingly, researchers must be aware of potential issues regarding intellectual property, including regarding patents, copyrights, and trade secrets, as issues involving all of these now play a role in collaborative research projects. It is increasingly possible to confront IP issues as researchers, and an awareness of both the law and the risks, as well as their relationships with the ethos of science can help researchers navigate potential problems.

4.3 Who Owns What?

Typically, and without the intervention of employment contracts, the authors or creators of a work are deemed the "owners" of intellectual property. Unlike ordinary ownership, which implies the ability to control completely the thing owned, IP owners merely have a restrictive right to prevent other from duplicating the objects of their IP rights, not to control the actual instances or copies. In other words, my copyright of a book give me no right to possession of all copies of the book. The same is true for a patent. It gives me no right to own all iPhones if I hold the patent to iPhones, but only to restrict the duplication of them. Things become much more complicated when one is not a sole inventor or author, and especially so when there is an employer.

Because most researchers are not working solo, and generally work within some institution, local laws and contracts will govern the allocation of IP ownership rights. Most employment contracts give the employer primary ownership of inventions of employees, where employees are paid for and expected to discover and create new and useful things. Because these relations are matters of contract, researchers taking positions in which it is possible that some research will lead to a patentable invention should read carefully their employment contracts to know for certain whether they will be named as an inventor in case of a new and useful invention or discovery. Employment contracts are often negotiable and researchers who wish to share in an invention, either through being named as an inventor or through profit sharing, should make themselves familiar with the applicable laws (which may prevent or enable such sharing) as well as work out the terms with their employers before they begin their research. Inventors working for institutions, including both private and public facilities, should not expect to claim ownership for their discoveries and inventions made while in the course of their employment. It is the norm for employers to claim ownership. Moreover, as partnerships among both private and public institutions increasingly become the norm, contractual assignments of IP rights have become more complex, and are part of the incentive created between funders and researchers. Granting agencies both public and private often allocate IP rights contractually as part of the funding instruments. Some public funders actually insist on open innovation, preventing anyone from claiming IP rights. All of these variations require careful attention so that once some potentially profitable invention results from research, no one is surprised by the claims or lack thereof to the IP produced.

Similar caution and discussion about authorship and copyright must be taken prior to the production of some written work or other copyrightable result from research. Although the author for copyright purposes is typically the person who creates an expression, this is complicated in research programs due to contracts, as with patents, as well as joint authorship. Grants, funders, institutions, and other collaborators may well have contractual bases for co-authorship, and thus royalties and other rights flowing from holding the copyright on some expression. Typically, copyright law does not protect data, although some jurisdictions have created special rights to protect it. Copyright generally only extends to original expressions, and the underlying data is discovered through observation and so not an original expression. The original expression consists in the manner of publication of the data, as well as the words, tables, etc., that are used to represent it. Again, as with all of the preceding, careful attention to contractual and local legal obligations should be paid prior to engaging in a joint research projects so that there are no surprises.

4.4 Not Treading on IP During Research

In order to insulate themselves from violations of IP, researchers must be aware of the nature of IP rights as discussed above, as well as the existence of certain exemptions. Until recently, there was no general exemption from IP infringement for research conducted in a lab (in the USA), although courts had in various jurisdictions and times made such and exemption. In other words, researchers who were simply doing basic science that used through duplication some patented product or process, would still be guilty of patent infringement even though they had no intent to profit from their research. By way of contrast, Article 22 of the Mexican Industrial Property Law states: 'The right conferred by a patent shall not have any effect against: a third party who, in the private or academic sphere and for non-commercial purposes, engages in scientific or technological research activities for purely experimental, testing or teaching purposes, and to that end manufactures or uses a product or a process identical to the one patented...' This exemption functions like that of numerous other national exemptions, allowing researchers to duplicate a patented process or product for purely scientific reasons without fear of infringement. It is noteworthy that the US does not have such a broad exemption, so when conducting joint research with US partners one must be careful regarding the locality in which the exempted research is conducted.

Fair use rights make complete control over expressions impossible. Thus, a copyright holder cannot prevent another author from excerpting or quoting from his or her paper. Fair use means that others are free to quote and excerpt the work of another to some limited extent, including especially for fair comment, refutation, critique, and parody. Authors who use the words or expressions of others under a claim of fair use must be mindful of course to properly attribute their source lest they find themselves guilty of plagiarism. Copyright may also cover presentations and other expository materials, including lectures and talks about a scientific program or project, if those materials become fixed in some tangible medium (e.g., a slide show, a recorded lectures, etc.) and so use of such materials without permission and attribution would ordinarily violate copyright.

Other legal exceptions may exist: for instance certain whistleblower protections may exempt one from claims of violating IP in case the disclosure is related to exposing some fraud or other misconduct covered by whistle-blower laws. However, absent such laws, researchers need to be aware of when and to what extent they may infringe a patent or copyright. Moreover, contract law may apply to trade secrets even where no patent or copyright exists. If one is working with trade secrets, it is likely that a "non disclosure" agreement was signed before the work started, and so if one is doing research under a NDA, one needs to be aware that breach of such an agreement can result in civil lawsuits and penalties.

4.5 Science and Competition

Consistent with what we have discussed above, specifically regarding the ethos of science, intellectual property issues relating to scientific research should not hinder the dissemination and replication (or falsification) of basic, scientific research. Without openness, science and its institutions cannot conform to communalism and organized skepticism. Inquiry into nature must be viewed as a joint effort of all scientists and their institutions, even as individual researchers and their programs compete for academic and scientific notoriety. Such competition fails the ethos of science if disclosure about discoveries is hindered inappropriately. This does not imply that all competition opposes the ethos of science, and it is arguably true that competition to some extent advances its ethos.

Researchers are not unethical by virtue of attempting to "scoop" another researcher or group. It is perfectly natural and generally consistent with the ethos of science for individuals and groups to wish to and attempt to beat others in reaching some goal. Where this becomes inimical to science is where the general progress of a scientific research program suffers due to such competition. One way it can do so is when disagreements about ownership, priority of discovery, or other conflicts end up resolved not through scientific means, but through legal redress. Scientific disputes about priority are as old as science itself. Consider the long disagreement and drawn out public dispute between Leibniz and Newton regarding the invention (or discovery) of calculus. Over time, and through the processes of scientific inquiry, this dispute has been resolved (they both did so, independently). What might have happened had either been able to claim copyright or patent over calculus? Would science have been served by a legal dispute, or would it have been hindered?

The ethical priority of scientists embracing the ethos of science, it would seem, ought to be the truth as we have discussed before. Secondary considerations such as profits, ownership, careers, etc., all follow the priority of truth and should there be some conflict, the ethos of science should prevail. Where competition spurs and does not hinder the quest for truth, then it ought to be encouraged and embraced. Where the truth becomes secondary, researchers ought to be wary. Indeed, wherever possible, openness should be the primary value embraced, even where profit and priority are sacrificed. Although laws allow for monopolies over new and useful products and processes, they are not required legally, and they may violate the ethos of science if they tie up the domains of science, as they have in the recent past.

Consider the recent controversy regarding the genetic mutations responsible for a large number of breast and ovarian cancers, the so-called BRCA1 and 2 mutations. When these mutations were discovered, scientists around the globe were working to find them, focusing on chromosomes that had been identified, published about, and considered to be the general location of important mutations.

4.6 Nature vs. Artifacts: What Ought to Be Monopolized Consistent with the Ethos of Science?

The intention behind intellectual property law has generally be considered: to create an incentive for creativity and invention. Basic science typically focuses on discovery, and as we have discussed above, scientists ought to concern themselves primarily with investigation of nature and her laws. The law of intellectual property typically attempts to protect the proper domain of science from that of technology, which can legitimately and presumably efficiently hold monopolies over inventions without impeding science. But this is not always the case, and sometimes the law takes a while to catch up with changing technology.

Consider the Myriad case. In the 1990s, numerous research groups around the world were honing in on the specific genetic mutations thought to be involved with high incidence of breast and ovarian cancers. The BRCA1 and 2 genes were finally identified as the site of a common mutation with a high correlation of breast and ovarian cancers, and among the researcher to identify the mutations was a group at the University of Utah, including lead researcher Mark Skolnick. A number of other researchers, including Mary Claire-King at the University of Washington, were very close to identifying the same mutations, and indeed identified related mutations. But when Skolnick's group found BRCA1 and 2, they obtained patents for them. The patents covered the unaltered, but merely isolated sequences associated with the genetic mutations involved in a high number of breast and ovarian cancers, and once obtained allowed Skolnick's private company (which licensed the patents from the University of Utah, where he also worked) to monopolize the market for genetic testing for the BRCA1 and 2 mutations. Patented technologies, developed in conjunction with basic science conducted with public money at public institutions, are common and not necessarily contrary to the ethos of science. But what if a patent issues too far "upstream?"

In the case of the Myriad patents, researchers around the world complained that when they sought to investigate the patented genes they would receive letters from Myriad threatening them with lawsuits. Indeed, under US law, researchers were not allowed to replicate the patented genes without specific license from the patent holder, although the research exemption discussed above applied to research centers in most other countries. For about 18 years, Myriad's patents on these genes allowed it to keep any other company out of marketplace for testing for BRCA1 and 2, but it also seems to have stifled some basic science. In was not until 2013 that finally a legal challenge that went all the way to the highest US court stopped the practice, with the Supreme Court holding that the patents were really on natural products and not on something man-made.

The legal history that describes how Myriad came to own patents on natural products is long and interesting, but it is beside the point for the purposes of our discussion. What is interesting for us is the line between nature and "not-nature" or "artifacts." If science must be able to pursue the truth, can monopolies be granted on natural products and phenomena without potentially impeding that search? The courts eventually reasoned that because the genetic sequences that were patented were not created but rather found and then "isolated," the patents covered something that was not "man-made." For similar reasons, one could not monopolize the theory of relativity, Newton's laws of thermodynamics, the element "oxygen," etc. these are all devised not by human inventiveness, but discovered through empirical research to exist in nature.

Perhaps one way to distinguish nature from artifact is to ask: does the thing exist only due to some combined human intention and design? In the case of natural laws and products, they exist despite all human intention and design, as do the mutations to BRCA1 and 2. Natural evolution brought them into being as they are, even though at some point some human "extracted" them from their surroundings and "isolated" them. If we wish to incentivize creation, then we should limit our grants of monopolies to human creations, but extending those monopolies to things that are not human creations, but that come to be discovered, stretches the idea of intellectual property too far.

The incentives for scientific discovery are built into the institutions and ethos of science itself. The desire to understand better the nature of the universe and its mechanisms drives scientific pursuit without, generally, the promise of economic reward. There are certainly other incentives attached to the various modern institutions of science, including notoriety, career advancement, the recognition of peers, etc., and these too function to prod the investigation of nature forward without the need for monopolies over the objects of scientific study. In fact, there is a general trend to try to open up science, recognizing that certain monopolies, including in publishing and through commercialization, may tend to undermine the foundations of scientific process.

4.7 Open Science as an Alternative

As an alternative to monopolistic tendencies, both through secret keeping and legal mechanisms, some embrace an "open" approach to all scientific inquiry, shunning trade secrets, copyrights, and patents in the process.

As we have seen in previous chapters, a lack of transparency either leads to or hides numerous instances of wrongdoing in science. Embracing the elements of the scientific ethos that have to do with communalism and universalism, as well as organized skepticism, all means at some level being completely open with one's methods and data. In order for science to function according to the Mertonian ideal, attempts to monopolize at any level will at some point defy the ethos of science, and presumably hinder progress toward the truth. An alternative approach is to be completely open with one's pursuits.

"Pay-walls", for instance, that help ensure profits for scientific publishers, arguably hinder science. Only researchers with institutional subscriptions to the large scientific publishers erecting those pay-walls will be able to read the papers published behind them. In order to test the claims in those articles, a toll must be paid that not every researcher or institution can afford. Publishing in "open" forums would be preferable to the aims and ethos of science as a greater audience of peers able to challenge the results published would be able to meet their scientific responsibilities. There would be an increased likelihood of either confirmation or falsification, and science would move forward more quickly. Norms and expectations regarding "impact factors" attached to subscription-only publications are a pressure against this, but if scientists and their institutions wish to conform better to the ethos of science itself, they will push back against these norms and expectations wherever possible.

The products too of basic scientific inquiry ought not to be monopolized as well if we embrace an "open" approach to science. One could conceivable reconfigure the nature of practical inquiry so that greater cooperation would emerge, reducing incentives to hide behind trade secrets or bottle up the results of our inquiries behind monopolies. It would be a major cultural and economic shift from the current system, requiring new incentives and influences that would drive science forward. One part of that might be better, personal appreciation for the nature of science in general, its ethos, and the responsibilities of individual scientists toward these.

None of this utopian "open" vision of science need preclude profits, technology, or capitalism. Rather, a respect for the divisions between basic scientific inquiry, and technological development, might help to differentiate and possibly help prevent the sorts of conflicts of interest we will discuss in the next chapter. There is no reason to think a totally open approach to discovering nature's truth would or should prevent the profiting by some over the uses of those discoveries in new creative activities. Indeed, science and technology have worked this way since the Enlightenment, and notable scientists were also inventors, creating new tools, objects, and processes as the result of their basic research, sometime also profiting thereby.

4.8 Conclusions

Intellectual property is a prevalent force in modern science and technology, and researchers will doubtless have opportunities to engage in ethical dilemmas potentially raised by these legal tools. Understanding their history, purposes, and natures, as well as various ways in which they can conflict with the ethos of science may be helpful in avoiding ethical harms. At the very least, scientists should be mindful of the duties, both moral and contractual, that arise with the publication and creation of the products of scientific inquiry, as well as the expectations that come from various forms of employment and other relations among scientists, their institutions, funding agencies, and governments. As with other matter discussed in this book, transparency and clear communication about roles and responsibilities early on can go a long way in preventing disagreements and potential harms that can arise due to intellectual property rights in a modern, complicated scientific environment. All of the various stakeholders should become familiar with their obligations, both ethical and legal before engaging in their creative activities so that what is owned by whom is clear before the work begins.

Study and Discussion Questions

1. What counts as "intellectual property" and what does not? What distinguishes inventions from discoveries?
2. Which values suggest a primary role of science over intellectual property rights? How do the Mertonian norms affect your consideration?
3. How much human intervention is necessary to turn something into intellectual property?
4. How do copyrights affect the progress of science, and how can we better accommodate science and maintain our interest in rewarding authors?

References

Biagioli, Mario, and Peter Galison. 2014. *Scientific authorship: Credit and intellectual property in science*. New York: Routledge.
Etzkowitz, Henry, and Andrew Webster. 1995. Science as intellectual property. In *Handbook of science and technology studies*, 480–505. Thousand Oaks: Sage.
Jensen, Kyle, and Fiona Murray. 2005. Intellectual property landscape of the human genome. *Science(Washington)* 310(5746): 239–240.
Kieff, F. Scott. 2000. Facilitating scientific research: Intellectual property rights and the norms of science – A response to Rai and Eisenberg. *Orthwestern University Law Review* 95: 691.
Koepsell, David R. 2003. *The ontology of cyberspace: Philosophy, law, and the future of intellectual property*. Chicago: Open Court Publishing.
Koepsell, David. 2011. Things in themselves: A prolegomenon to redefining intellectual property in the nano-age. *Journal of Information Ethics* 20(1): 12.
———. 2015. *Who owns you: Science, innovation, and the gene patent wars*. Malden: Wiley.
Long, Pamela O. 1991. Invention, authorship, "intellectual property," and the origin of patents: Notes toward a conceptual history. *Technology and culture* 32: 846–884.
May, Christopher, and Susan K. Sell. 2006. *Intellectual property rights: A critical history*. Boulder: Lynne Rienner Publishers.
Nelkin, Dorothy. 1984. *Science as intellectual property: Who controls research? AAAS series on issues in science and technology*. New York: Macmillan Publishing Co.

Prager, Frank D. 1944. History of Intellectual Property from 1545 to 1787, A. *Journal of the Patent and Trademark Office Society* 26: 711.

Schoen, Robin A., Mary E. Mogee, and Mitchel B. Wallerstein (eds.). 1993. *Global dimensions of intellectual property rights in science and technology*. Washington, DC: National Academies Press.

Timmermann, Cristian. 2013. Life sciences, intellectual property regimes and global justice (excerpt). Diss. Wageningen University.

Chapter 5
Conflicts of Interest

Abstract There are many institutions and relationships that researchers are involved in every day. One's affiliation to a university, funding agencies, department, corporate partners, or even family and friends may present conflicts of duties that can impede research or cause harm. Sorting out the nature of our duties, and being aware of the various individuals and institutions to which we owe duties, is essential to avoiding the harms that may come. Not all conflicts of interest can be avoided, nor need they be harmful, and so recognizing when and how to avoid, or at least be transparent, about conflicts when they arise and prevent harm when possible is crucial. As with other norms of scientific and research behavior, we can be guided by the Mertonian norms to help navigate the dangers presented by complex collaborative ventures common in modern science, and devise mechanisms that may help us to avoid the harms that may accrue. In this chapter, I try to define a conflict of interest and provide some guidance and examples to help researchers understand them, avoid them, or at least be aware of them and provide the best transparency to the parties involved.

5.1 What Is a Conflict of Interest?

One way to begin to think about conflicts of interest in research is to consider what you think the central role of science, as a discipline, is supposed to be. Whatever that central role is, when other goals or motivations become present in a researcher, institution, or affiliated group which are not necessarily the same as that central role, or even worse, *conflict* with it, then there is a possibility that a particular research project may go astray, perhaps in harmful ways. The danger of harmful conflicts of interest is rising as the barriers between various institutions involved in the pursuit of basic knowledge become blurred.

The proper aim of science is developing a clearer understanding of nature. Through well-established methods of observation, hypothesis, testing, and building of theories, we gain an ever better understanding of the objects and processes that rule the universe. In order to conduct this sort of study properly, we need to be in a state of "equipoise," in which our emotions are not vested in a particular outcome other than discovering whether our hypotheses can be corroborated by evidence or falsified. There are countless examples in science of researchers who failed to properly detach themselves from a particular outcome and who made significant errors, sometimes costly to the entire institution of science, that only later came to be corrected. And while science is indeed a largely self-correcting institution, and errors do come to be corrected over time, when human subjects are involved the mistakes made by improper detachment can be costly not just in money, but in lives.

Maintaining a proper understanding of the role of science and the researcher in its institutions helps us to start to understand how conflicts of interest may arise, be recognized, and dealt with. The nature of science has never been as pure as the ideal stated above. For its continued progress in the world, it has depended upon interaction with human institutions other than science, including the market, states, churches, etc. those interactions sometimes have come with support, and at other times resistance. Other times, science and other institutions interact only weakly or not at all, but work for opposing or orthogonal goals. Just as often, researchers working within the institutions of science are also members of some of those other institutions, and may themselves experience differing constraints or impulses as a result of their membership in those institutions.

The challenge of scientists and the institutions that support them and their projects is to attempt to marshal the various goals and aims involved in all stakeholders, and help ensure that they are directed in a similar manner, toward similar goals, mindful of the potential pitfalls that may arise if interests conflict and are unattended to. Due to the increased intercourse among science and other institutions, necessitated in part by changing economic commitments both to basic research and to developing technologies, there has been a growth in both scholarship and practice in understanding, recognizing, and confronting potential conflicts of interest. Below I will examine conflicts of interest (COIs) in theory and in practice, offer some advice in spotting and dealing with them, and conclude with some examples both real and hypothetical.

5.2 The Interests of Science

Science is an evolving institution involving people in various fields generally interacting in ways that are socially constructed, bound by poorly define mechanisms, and interlinked with other institutions as mentioned above. The "rules" by which it operates are not centralized, not formalized, not overseen in any central manner, and evolve over time. As discussed in Chapter I, for science to work as it ought, to uncover over time the laws that govern the universe, the "ethos" of science demands certain types of behavior by participants in its institutions. Universalism, communalism, disinterestedness, and organized skepticism all suggest certain necessary behaviors, and all assume something about the nature of science and its aims. We assume a stance of scientific realism, in which the laws of nature govern all objects and processes everywhere, and which can be generalized about properly by scientists taking the proper stance of objectivity and disinterestedness.

5.2 The Interests of Science

Of course, institutions, especially when they are so unintentional, so organic and so distributed, do not have "interests" any more than they have "thoughts." People who involve themselves in doing science, however, ought to understand that in order for it to proceed properly, in the general direction of increasing our prediction, control, and understanding of nature and her mechanisms, certain principles must be followed and the general ethos of science must be abided by. Those who consider themselves scientists, or part of the institutions of science, should therefore be aware of their responsibilities in ensuring that science proceeds as smoothly as possible, which means keeping their personal interests and the interests of "science" distinct, but aligned.

Science moves by fits and starts, there are many dead-ends and u-turns, but it is by ascribing to the general ethos of science that those working within its institutions can help become assured that there will always somehow be progress. No one scientist can be responsible for that progress, but all those within science who are responsible, will help ensure that progress. This includes noting individual scientist's errors, limitations, and helping to correct those when they occur. It also means recognizing that science is conducted by humans, and understanding our flaws and tendencies toward interests that, when not properly aligned with the interests of science, may turn into conflicts.

The ongoing interest of the institutions of science must be, primarily: *truth*. This assumes that the language and models we use to try to understand the universe have some relation to the universe itself, and that the general position of scientific realism is itself true. There are bodies and relations among them that exist and abide by natural laws with or without scientists who try to uncover those laws. In order to achieve a better picture of the truth, scientists and others working in its institutions must place this value above all other values, which means sometimes that in order to achieve and respect the interests of science, we must set our other interests aside, or even sometimes subvert them when they conflict.

The potential for conflicts is clear. As humans, we have many immediate interests that have very little to do with and may even sometimes conflict with "the truth" as an overarching interest, including providing for ourselves and our families, our egos, our emotions, etc. It takes a certain amount of detachment to recognize this, and to note that in our daily lives there are many opportunities that present themselves to us to promote our own selfish interests that may be at the expense of both the interests of others, and of the "truth" as a grand notion. Indeed, much of our modern lives, and certainly a fair amount of the modern marketplace, depends upon appealing to our individual, sometimes selfish notions of personal wellbeing at the expense of the wellbeing of others. It is no different in scientific research, especially as often conducted in the realm of academic science. Because "science" is not itself a monolithic institution, the institutions in which scientists work also create pressures that affect individual scientist's interests. Ideally, all of those interests ought to align. That is to say, once every stakeholder in science understands that truth is for the benefit of all in the long term, then there should be no conflicts. However, science's time scale and differing time-scales for humans and their interests do not always coincide, and thus the definition of "benefit" may shift according to shorter-term interests than science allows for.

Below I examine how interests of individual scientists, organizations, and sometimes entire communities may come to be in conflict with the interests of science, examine some examples and potential solutions, and discuss ways to recognize, educate, and ameliorate conflicts before or when they emerge. Along the way we will look at the increasingly complicated milieu that has helped to generate new types of conflicts, and also examine the types of ethics committees, behaviors, and solutions that may exist for the purposes of resolving conflicts if and when they emerge. Of particular interest is the emergence of private ethics committees, which while public committees certainly also have had to deal with their own conflicts in some instances, offer the possibility of new types of conflicts that may affect not just researchers, but the institutions we have established ostensibly to help to prevent ethical lapses in the presence of conflicts of interest.

5.3 Interests of Scientists

At its core, science is conducted by individuals working in loose, often dispersed and certainly diverse communities. These communities may come to be defined by institutions, or they may only be defined by the subject matter of the study of those working in a particular domain. Scientists, just like members of any other profession, have interests outside of their work. Typically, they have families, jobs, friends, and activities they pursue that have little or nothing to do with the general goal of science. Also typically, those interests tend not to conflict, even when they may diverge or have essentially nothing to do with the goal of science.

Over time, and as science has professionalized both within and outside academic institutions, the possibility for conflicts among personal interests and those of science has increased, sometimes in ways that may be difficult to spot. The career, ego, notoriety, or fame of a particular scientist always had the ability to create some conflict, but when attached to other offices and incentives, the effects may become compounded.

Of course, properly aligned with interests in fame, careers, ego, etc., the scientist should think of these interests in the long run, realize that they are all dependent upon the proper pursuit of science, and that failing to abide by the ethos of science in general runs a risk of undermining personal interests as well. History is replete with examples of short term gains through violations of science's ethos, at the long term expense of the reputation, wealth, etc., of individual scientists. Eventually, the truth will come out in the conduct of proper research. Somewhere, someone at some point will either reproduce a result or fail to do so, leading to the scientific community either coming to discover an experimental error, or at the very worse, a falsification of some sort. Retractions or worse only serve to undermine a person's stature in the scientific community, and typically overshadow the short term gains made though conscious violations of norms. The cure for the possibility of apparent conflicts between interests of individual scientists and those of science in general may include increased awareness of the consequences over the long term for bad behaviors that conflict with science's ethos.

Beyond the individual and immediate interests of scientists are also their interests in the profession, as members of it and as its beneficiaries, both immediate and distant. Indeed we are all beneficiaries of science, but as with other professions (lawyers, doctors, etc.) the public perception of the profession is altered both positively and negatively by the conduct of individual scientists, especially and unfortunately when a member of the profession does something visibly harmful. Conversely, and as an incentive for good behavior, as the collective view of a profession improves, so does each individual member of that profession benefit. Science is conducted largely through the good will of society, which serves to fund much of basic science, and which expects a certain level of conduct from its practitioners. Unlike doctors and lawyers, the level of remove between scientists and the public is often greater, and so many laypersons have very little knowledge of what scientists are doing on a daily basis. This means that too often members of the public only hear about science in two instances: when something very exciting happens and is reported upon by the media, and when something rather bad occurs. Again, the interests of individual scientists do align with those of the profession, and with the ethos of science, and suggest abiding by certain norms, and short term judgments about those interests may result in conflicts leading to bad behaviors.

Scientists must be aware of the nature of their immediate and long term interests. Examples such as those I will discuss briefly show that failing to align one's interests with those of science may have deleterious consequences both for the individual scientist, and for the domain of science and its institutions. Below I will examine how institutions may help create and also avoid conflicts, and the means by which science depends upon and may come to be influenced by those institutions.

5.4 Other Institutional Interests

Science is conducted by scientists generally working within some institutional setting. The notion of the lone scientist, working at home in an attic lab is no longer viable, if it ever was. Rather, most basic science is conducted in academic and research institutions, such as private and public universities. Universities have increasingly complex webs of relationships. Increasingly, there are new forms of partnerships among universities, corporations, independent research and development organizations, privately funded and publicly funded cooperative ventures, all with their own sets of interests including basic research as well as various others. Universities may well differ in their interests one from another, with private and public universities having to answer to different sets of coordinators and experts. Private research facilities too may have their own sets of interests, including investors and shareholders, public relations concerns, and contracts with yet others that may at any time be affected by a particular study. Corporations owe their primary, financial duties to their shareholders and investors, corporate venture capitalists, and their boards of directors. Science may well, and certainly ought to, be part of their concerns, but maintaining the primacy of the interest of science and its ethos can be difficult given other, pressing short-term considerations.

Because of the nature of what we could call "hosting" or "funding" institutions, scientists may feel pressures that sometimes are not aligned well, or may indeed conflict with, the interests and ethos of science. A particular study which does not produce the result expected and desired, may have short-term deleterious effects on the host or funder. Expectations may be delayed or foiled, and the public perception of the host or funder may suffer. Part of the role of hosts and funders must thus be to keep the interests of science in perspective, to recognize that despite the other interests pressing upon them, their general approach to science, in order not to suffer the public and scientific harms associated with scientific error, must coincide with its ethos. For all concerned, this means that they must adopt a position of "equipoise" regardless of their other interests.

5.5 Equipoise: A Duty of Scientists and Their Institutions

Having in mind and as a primary interest the ethos of science, and adopting a perspective of equipoise, means that we recognize that science proceeds in a certain manner, and that our outlooks about it can affect its proper progress. Every study should be conducted without assumption about its outcome. Science has historically gone down many dead ends when researchers, having in mind and desiring a certain outcome, erroneously interpret data such that it is more favorable to their expectations. Failures of equipoise can occur rather innocently, spurred by natural, human emotions such as hope and excitement, and need not be nefarious or otherwise badly intended.

When we expect certain results, we may be inclined to see them where they do not clearly exist. One prime example of this is the "N-rays" controversy of the late 19th century. Shortly after the discovery of X-rays, Prosper-René Blondlot was conducting experiments on electromagnetic radiation, specifically attempting to observe polarized X-rays, when he noticed apparently changes in the brightness of photographed sparks in an x-ray beam. Unable to account for the changed brightness according to current knowledge about X-rays, he proposed that a new form of radiation was being observed which he called N-rays. Using "detectors" made of a dim phosphorescent material, researchers around the world began perceiving N-rays, concluding similarly that unexpected increases in luminosity that they perceived were corroboration of the new form of radiation. Within a few years, over 300 paper were published on the subject. When some rather notable scientists, including Lord Kelvin, failed to be able to reproduce the results, doubt about the existence of N-rays began to grow. Only when Robert Wood visited Blondlot's lab, and surreptitiously replaced a part of the apparatus and test materials during the course of the experiment, and Blondlot noted no change in his detection of N-rays, did it become clear that what was being perceived did not exist. It was the hopeful observation of a researcher who, like several hundred others around the globe, was seeing something he wanted to see. He lacked equipoise.

Another term for the lack of equipoise is "experimenter bias" and studies are often designed in such a way as to minimize this, as with double-blind studies where the experimenter is supposed not to be able to know which is a test material and which is a placebo or control. But whole studies cannot be fully protected in the same way from unknown and unwilling bias, even less so from intentional bias. Experimenters necessarily work knowing from whom their support comes, and the general aims of their research. Equipoise is already fragile, as the N-rays debacle shows. We can easily fool ourselves even when we don't have some other interests than our own interests in discovery. When, however, other interests such as continued funding support, keeping our hosts and funders happy, as well as our own interests in job security and even fame enter the equation, maintaining equipoise can become especially imperiled.

Maintaining equipoise is consistent with the Mertonian ethos of "disinterestedness," which means that we should have no expectation of "success" in our studies other than to observe something. The null hypothesis, which is the idea that there is no causal connection between two observed phenomena, ought to be the default expectation. Only through continued, challenging testing and further confirmation should we be moved from the null hypothesis toward a theory. Science must then necessarily be slow, perhaps frustratingly so for researchers, their hosts, and their funders, but pathological instances of experimenter bias are more harmful to the institutions of science and its participants over the long term.

5.6 The Problem of Private Ethics Committees and Contract Research Organizations

Ethics committees are now mandated by international treaty and national laws created in the wake of a number of very public and tragic failures to abide by basic ethical principles in human subjects research. Any study that is conducted using human subjects must be first approved by an ethics committee applying the core principles of bioethics enunciated in such documents as the Belmont Report and adopted in the Treaty of Helsinki. In most countries, this means that ethics committees must be established at research and academic institutions at which a majority of research using human subjects occurs. Until recently, publicly created and funded IRBs or Ethics Committees were formed, made of community volunteers, peers at those research universities, were the norm. More recently, publicly-funded and run ethics committees are perceived to be unable to meet demand and progress stymied as a result. Thus the growth of so-called "private IRBs," often in conjunction with the growth of "contract research organizations" and in-part necessitated by the changing nature of scientific research in a technological milieu.

Because a fair amount of research is now conducted by private research and development teams, working often in private corporations, the incentives and concerns of the developments of that research is different than publicly-funded research. Moreover, with the increase in "private-public partnerships," consisting of cooperative ventures between public and private institutions, a complex web of interests and fears have helped to spur the development of contract research organizations which can take portions of a research project, protected by contractual relationships like "secrecy" and help to prevent financial losses or "thefts" while commercializing a technology. Because of associated concerns over secrecy and privacy, using private ethics committees might well be favored in such arrangements.

In most countries, ethics committees are mandated by law, and overseen by some governmental organization. Often, this means their operations are periodically reviewed and assessed, and that they can be sanctioned for poor results or failure to apply proper standards. In some cases they may be shut down and reformed, and they serve at the behest of the public and the public entities that provide their oversight. Private ethics committees often lack the same sort of transparency as public ones, which is part of the allure of their use in corporate and private-public partnership research. It also means that the same sort of incentives and potential pitfalls do not exist. Already, public ethics committees may well, even while monitoring studies for the existence of potential conflicts of interest, have conflicts of their own. When private, and contracted-for, the nature of those conflicts may well grow.

Private ethics committees are often hired when a company needs some assurance of speedy review as well as need for protection of secrets. The duties of the private review committee then seem to include a speedy review, as well as non-disclosure. But these values may each conflict with the proper progress of science in accord with its ethos. To whom does the private ethics committee owe which duty, and how does its duties to both science and the public mitigate against its fiduciary interests in its contracts?

I will look below at a very famous case that offers us some ability to see how conflicts may exist, may even become deadly, and how ethics committees may err for them, then we will look again at how private and public ethics committees may help avoid COIs.

5.7 The Gelsinger Case

One case that gravely highlights the potential pitfalls when one or more conflicts of interest are present is that of Jesse Gelsinger, an early human subject in a gene therapy trial at the University of Pennsylvania. In the early 1990s, some minor success in proving the possibility of gene therapy, by which damaged genes are replaced in vivo with functioning genes, spurred a race to develop successful and potentially life-saving gene therapies for a variety of diseases. Gelsinger had ornithine transcarbamoylase (OTC) deficiency, a metabolic disorder that affects 1 in 40,000 children by impeding the elimination of ammonia. It is a monogenic disease, meaning that it results from a mutation in one gene only, making it a perfect target for gene therapy, which seeks to use a vector such as a virus to deliver a non-mutated string of nucleotides into the cells of a living subject to replace a malfunctioning, mutated string. If Jesse Gelsinger's mutation, which was only partial and had allowed him to survive for nearly two decades rather than the usual five years as long as he kept to a very restricted diet.

In 1999, Jesse Gelsinger turned 18 at the same time that a human trial of gene therapy to repair OTC mutations was occurring, headed by James Wilson. Without getting into the details of the trial and its failures, it is instructive to note the various relationships among the stakeholders as illustrative of the complex nature of modern science and the potentials for conflicts of interests to not only be created and even recognized, but to cause failures at nearly every level of review. In this case, the failure was fatal to 18 year old Jesses Gelsinger.

In 1993, James Wilson joined the faculty at the University of Pennsylvania with both a promising track record in research on gene therapy, and a for-profit company, Genovo, that owned several of Wilson's patents on gene therapy-related processes. Soon after joining the faculty, Genovo won $36 million USD in venture capital funding, after which the University of Pennsylvania eased its existing conflicts of interest guidelines to allow Wilson to keep a higher percentage of the profits of his company than normally allowed. When Wilson came to U. Penn (Penn), it was with the promise that he would head there a university-hosted institute, the Institute for Human Gene Therapy (IHGT). Genovo was a major funder of the IHGT. Penn also came to hold shares in Wilson's for-profit company, Genovo, and thus would potentially build upon its endowment just as Harvard had with its co-ownership of patents on the Oncomouse. Meanwhile, Wilson remained an active participant in Genovo, the head of the Institute for Human Gene Therapy, a faculty member at Penn, and the principal investigator on the OTC human gene therapy trial which had moved from animals to humans by the time Jesses Gelsinger was old enough to consent to become a subject in the trial. Wilson had foregone a salary from Genovo, but held more shares than normally allowed by a faculty member thanks to Penn making an exception to its conflicts of interest guidelines for Wilson. Even so, two separate committees were established at Penn to oversee Wilson and his various relationships and research, in order to try to help mitigate any effects from conflicting interests. The members of those committees, as with the ethics committee that oversaw the research, were peers at Penn.

Indeed, the committees noted various concerns from the relationships of the school, the company, and the institute, including the potential for profit from the success of the study due to both Wilson and the former dean of the medical school's and head of the university's health system ownership of patents in the gene therapy processes at the heart of the study. Because of the various concerns, and the troubling nature of the relationships in this "private-public partnership," lawyers representing Wilson, Genovo, the IHGT, and Penn worked out a variety of agreements about profits, liabilities, and in general attempted to alleviate the appearance of conflicts among the parties so as to help prevent financial interests from interfering with the study. In any event, the various parties all stood to profit not just scientifically, but financially from a successful study, perhaps significantly. Moreover, as part of the consent procedure, subjects of the study were advised of all of these relationships and the potential for conflicts of interest.

A few days after an initial injection with altered adenoviruses containing the corrected nucleotide strings meant to repair the mutated OTC genes in his liver, Jesse Gelsinger died of an acute immunological response. After the death, the FDA halted the trial, and conducted a review to determine the cause of the fatality. Its review found 18 specific failures by the team at Penn, including: admission of participants without documenting their suitability (some were apparently too sick to be properly admitted to the study); failures to properly advise subjects about the risks; omission of data about animal deaths during animal testing in the consent form; improper signature gathering; failure to properly disclose human subject side-effects observed in other subjects; failure to properly track health of human subjects during the course of the study; improper custodianship of experimental materials, among other failures. The FDA also raised questions regarding the nature of the relationships and their inherent potential conflicts of interest.

Today, the Gelsinger case is often cited for its illustration of how conflicts of interest, especially in an evolving research milieu in which universities and corporations do business and science together, may evolve and lead research astray. While Penn and Wilson and other stakeholders insist that their guideless and oversight were not the proximate cause of the death of Jesses Gelsinger, it is certainly worth noting that the university had already waived some of its conflicts of interest guidelines in establishing its relationship with Wilson, the IHGT, and Genovo. It is also certainly conceivable that all parties were in part influenced by the potential for profit, both monetary and otherwise (fame, etc.), and were not properly focused on the ethos of science, and did not sit in the proper position of equipoise.

5.8 What Can Be Done?

As the Gelsinger case so aptly demonstrates, mere awareness and disclosure of potential conflicts of interest may still not prevent harm. Increasingly, journals and funding agencies are demanding transparency about interests that may exist and conflict with the primary duty toward truth in scientific research, but again, this may not always prevent harm. How can we guarantee that researchers remain focused on the ethos of science, sit in proper equipoise, and avoid allowing other interests from intervening unethically? It could well be that no institution can do so adequately, and the evolving nature of science and its forms of funding are not likely to make it easier to avoid the creation, influence, and appearance of conflicts.

The focus must ultimately fall upon the individual researcher. When we recognize that we are all susceptible to various influences, even mundane and perfectly understandable needs for security in our incomes and our jobs, not to mention careers, prestige, profits, and other natural and common interests, then we should be aware that we can lose sight of the interests of science itself. Both private and public ethics committees too must be mindful that they are a part of a larger, amorphous, and important institution called science, and its ethos, the advancement of human knowledge through the development of better theories describing nature's parts and processes, is the primary concern of every scientific endeavor.

Of course, meanwhile institutions should be vigilant, even as each member of those institutions must be aware and on guard against the potential harms from conflicting interests. Transparency is certainly an important tool, but it is not the final end or goal. Even with transparency, actors may not always act according to the proper aims of science, and may allow intervening interests to cloud the pursuit of the truth. At every level, from the individual researcher to oversight bodies, to society, we must focus not on the profitable nature of scientific discovery (as indeed we do constantly profit from it) but rather upon the aesthetic ends of science, for lack of a better description. When truth and discovery are seen as valuable ends, and dead-ends, failures, and other consequences that don't make headlines are seen as a natural part of the scientific process, then a shift of focus toward the interests of science itself ought to help prevent us from becoming side-tracked, becoming personally vested in some particular outcome, and even harming others through improper focus on the wrong interests as we seek the aims of science.

Study and Discussion Questions

1. What is a conflict of interest? Which interests may collide rather than cooperate? What is the proper first interest of a scientist?
2. How has modern science exacerbated the possibility and effects of conflicts of interest? How might we try to alleviate them again?
3. How do conflicts of interest effect equipoise, and why is a lack of equipoise a threat to the norms of science?
4. How might you have helped alleviate the conflicts of interest in the Gelsinger case? What practical steps could have been taken?

References

Ascoli, Marcel, and N. Les Rayons. 1977. Rene BLONDLOT N-Rays. *American Journal of Physics* 45(3): 281–284.

Ashcroft, Richard. 1999. Equipoise, knowledge and ethics in clinical research and practice. *Bioethics* 13: 314–326.

Barber, Theodore X., and Maurice J. Silver. 1968. Fact, fiction, and the experimenter bias effect. *Psychological Bulletin* 70(6p2): 1.

Bekelman, Justin E., Yan Li, and Cary P. Gross. 2003. Scope and impact of financial conflicts of interest in biomedical research: A systematic review. *Jama* 289(4): 454–465.

Freedman, Benjamin. 1987. Equipoise and the ethics of clinical research. *New England journal of medicine* 317(3): 141–145.

Kintz, B.L., et al. 1965. The experimenter effect. *Psychological Bulletin* 63(4): 223.

Klotz, Irving M. 1980. The N-ray affair. *Scientific American* 242(5): 122–131.

Kubiak, Cinead R. 2004. Conflicting interests & (and) conflicting laws: Re-aligning the purpose and practice of research ethics committees. *Brooklyn Journal of International Law* 30: 759.

Newton, Roger G. 1997. *The truth of science: Physical theories and reality*. Cambridge: Harvard University Press.

Nye, Mary Jo. 1980. N-rays: An episode in the history and psychology of science. *Historical Studies in the Physical Sciences* 11: 125–156.

Relman, Arnold S. 1985. Dealing with conflicts of interest. *New England Journal of Medicine* 313(12): 749–751.

Resnik, David. 2004. Disclosing conflicts of interest to research subjects: An ethical and legal analysis. *Accountability in Research: Policies and Quality Assurance* 11(2): 141–159.

Rodwin, Marc A. 1993. Medicine, money, and morals: physicians' conflicts of interest. New York: Oxford University Press.

Shrebnivas, Satya. 2000. Who killed Jesse Gelsinger? Ethical issues in human gene therapy. *Monash Bioethics Review* 19(3): 35–43.

Smith, Lynn, and Jacqueline Fowler Byers. 2002. Gene therapy in the post-Gelsinger era. *JONA'S Healthcare Law, Ethics and Regulation* 4(4): 104–110.

Steinbrook, Robert. 2008. *The Gelsinger case*. New York: Oxford University Press.

———. 2011. The Gelsinger case. In *The Oxford textbook of clinical research ethics*, 110–120. Oxford: Oxford University Press.

Chapter 6
Autonomy, Dignity, Beneficence, and Justice

Abstract Much of our concern in the field of applied ethics has to do with the central principles of modern medical ethics. As it turns out, the ethos of science also demands that we abide by these same principles. At the outset of this book, I described the emergence of modern applied ethics, its origins in philosophical ethical theory, and the development of post-Nuremberg principles, codes, and institutions. Now we will explore a bit more in depth the nature of these major ethical principles as applied to human subject research. Norms that have been largely developed after World War II demand that human subjects be treated according to certain, basic ethical principles, including: autonomy, dignity, beneficence, and justice. In this chapter I discuss these principles briefly, and provide an argument for their adherence to and emergence from norms of scientific behavior generally.

6.1 The Emergence of Medical Ethics

The ethical principles that lie at the center of medical ethics evolved over time in various ethical traditions, dating as far back as Ancient Greece. Now the "Belmont Principles" or Nuremberg Code are usually referred to when we are concerned about how to proceed with some research when it involves some human subjects. Typically, this has meant that if we are testing drugs or devices, surgical procedures or other interventions involving doctors and patients, we have to take certain ethical precautions. While this is now institutionalized through ethics committees, reviews, and the use of experts involved in checking the nature of a study before it begins, the same duties are also ongoing personal duties of those doing science. Respecting the rights of others is arguably a duty to which we are bound regardless of the creation of some agency or board to oversee us. This is part of the theoretical justification for the Nuremberg Code itself in the absence of some formal code or treaty. How was it that the Nuremberg judges found that there were "crimes against humanity" that exist despite the absence of international criminal law, specific treaties contemplating those crimes, or even a system of police and judges that can oversee it? In sum, the ethical duties they found to exist are rooted in our human practices, norms of behavior that relate to those practices, and perceptions of right and wrong that are independent of culture, institutions, or states.

Regardless of the specific ethical theory from which we might consider any one of these rights or duties to emanate, we can discuss them in terms of positive norms that are now generally accepted internationally, and that guide good research behavior in a wide range of fields, even outside medicine. Many of these same principles may be familiar to us in our everyday interactions with other people and our environment.

Consider, for instance, the notion of autonomy. Do we wish that others allow us to make our own decisions, especially about our most precious resources including our bodies and minds? Do we not only expect to be treated as though we have at least that much autonomy, but also treat others the same way? Dignity is something too that each of us has a right to -- to be treated as though we are equally entitled to respect regardless of our background or station, and to grant the same to others. We do not wish to be used as some means to some end, but rather expect to be treated as though we are persons, not instrumentalities. If we are consistent, we do the same for others. Beneficence too is something we expect and hopefully practice simply as part of being a member of a civil society. Our actions ought to be motivated by good intentions, we ought to be inspired by good will to at least cause no harm to others, and at best to do good. When we are beneficent, regardless of our internal motivations, we can generally expect societal rewards, and overall happiness can increase. As with many of these values, cases can be made for their usefulness even if we do not necessarily think their origins are natural or "deontological" (based on some duty). Finally we expect that we should be treated justly, and that our rewards and punishments are deserved, measured, and consistent with those of others. Justice demands that people be treated equally, afforded the same opportunities or in some cases provided means of redress of natural shortcomings, as well as equal access to goods and services. Again, as with other rights we expect for ourselves, when we are consistent we recognize these rights for others.

Without much in the way of case study or ethical theory, we can already begin to see how recognizing the above duties in our ordinary interactions can help us to deal with human subjects of our scientific studies in medicine and in other fields. If we owe these duties and expect reciprocity in other situations, why should we not do the same in our research? One reason that scientists may not automatically apply common moral principles to scientific research may be the general perception that utilitarian concerns outweigh other moral duties. For instance, the notion that the greater good often outweighs individual harms, and that science is always focused on the greater good. But even in our everyday interactions we do not generally think that the greater good justifies every sort of individual harm. We routinely act as though we and others have certain inviolable rights, such as to our life, liberty, and property, and that redistribution of these merely to enhance the common good would be somehow wrong. What makes science exempt from these considerations? Increasingly, we are realizing that nothing ought to even while in the past some scientists have acted as though science is immune from the same moral standards we hold ourselves and others to in our daily interactions.

So what special considerations are there for scientists when conducting their research in accordance with the values and principles implicit not just in the

Nuremberg Code, but in society as a whole? Scientists using human subjects, and perhaps also considering the impacts of their non-medical science on society in general, are actually in a special relation with others. There is a significant amount of trust placed in science and scientists, as well as a number of advantages. For instance, society funds science both directly and indirectly, and those who are driven by their passions for investigating nature's truths are afforded the opportunity to do so as well as the trust of those who fund them to do so in ways that do not harm. Scientists thus have a special responsibility, rooted in part in that trust, and based in part on their special relationship to the knowledge they are seeking and their insights into potential harms. Let us examine below a bit more in depth about the principles above and their relations to scientific experiments, clinical trials, and the institutions and duties of science and scientists.

6.2 Autonomy

The term "autonomy" comes from ancient Greek, *autonomia* (n.), *autonomos* (adj.) from *autos* – self, and *nomos* – law. Essentially, it means self-governing. It was not considered a virtue or value to be held by individuals, according to Greek philosophers. In fact, Plato's *Republic* makes clear that he considered democracy dangerous inasmuch as individuals were not capable or trustworthy for self-government. Autonomy pertained until recently to freedom for states, meaning sovereignty only for governments to direct their own affairs free from intrusion by other states to manage their own polities. Until modern times, personal autonomy was viewed with distrust, and with the advent of Christianity, one's will was to be subject to the will of God and thus the state which was typically conceived as the embodiment of God's will on earth through a divine sovereign. It was not until after the Reformation and with the rise of The Enlightenment that the notion of personal autonomy emerged as a critical concept and formed the basis for modern political liberalism.

After the various revolts and rejections of the Catholic Church in Europe, and the emergence of various Protestant sects besides Lutheranism and Calvinism, so too came a greater variety and freedom of political thought. Luther and Calvin both embraced notions of "free will" that played a central part in their theologies, applying the concept to the general freedom that God gives to us to make choices, although they embraced notions of pre-destination of the soul. Nonetheless, the revolt against Catholic States introduced the idea that individual freedom of religious conscience could outweigh the dictates of a state and state-sponsored religious dogma.

With the advent of The Enlightenment, and the emergence of philosophers like Immanuel Kant, John Locke, and John Stuart Mill, the notion of autonomy took a central place in what was to become the dominant form of modern political thought: liberalism. Locke's conception of autonomy regards mostly political life, and forms the basis for modern liberal revolutions, placing the consent of the governed and freedom of personal conscience as essential elements of a just polity, including respect for fundamental natural rights of life, liberty, and property. Kant's conception regards the nature of moral autonomy, not so much personal autonomy. Kant thought that the basis for the good is rooted in the good will -- the will that is directed toward that which is our duty. Through our capacity for deliberative reason and its proper exercise, we are able to determine and choose the good, and thus be good. Moral autonomy is the cornerstone for achieving the good for Kant. It is through this practice, and properly motivated that Kant views us as achieving our dignity as humans. This notion of moral autonomy is closely akin to what we might consider freedom of conscience, a notion that John Stuart Mill (who you will recall has a very different conception of morals based not on Kant's deontology but rather on utilitarianism), embraces but also expands upon significantly. Mill embraces a view of broad autonomy, limiting freedom of action only to that which violates the "harm principle." We are free to do with ourselves, including in our thoughts, actions, and speech, anything which does not cause harm to others. Mill's broad conception of autonomy is central to much of modern liberal thought, and is reflected in constitutions and legal systems, as well as international agreements, treaties and charters.

The major philosophical notion opposing the notion of autonomy is "paternalism," or the notion that someone in some position of authority over another has the ability and even right to coerce or prevent the other from acting, or has the ability to act toward them in a way that violates their will. Clearly, paternalism was the dominant philosophy behind the pre-modern sovereignty of states (and churches) over individuals. In modern medical ethics, some mix of autonomy and paternalism tends to dominate, a conception we might call "procedural" autonomy that recognizes that the freedom of individuals to choose rationally, to govern their own body and conscious responsibly, may differ from one individual to another, and within one individual over time. For instance, as children we do not have the necessary tools to make free choices using our inherent dignity and rationality. We are limited by our capacities and so too will our "free" choices be actually limited. Thus we accept that parents act "paternalistically" toward their children without viewing that as a moral wrong. This is the proper role of a parent, to help their children develop the capacity to be rational, free agents, while restricting their choices in the process of that development. But parental paternalism is expected to dissolve over time, and to be replaced with the rational free agency of their children as they grow.

Physicians (and scientists in other fields interacting with human subjects) stand in a special relation to the knowledge they have about the science they are employing as a parent may be to their years of experience in directing the agency of a child as they grow. Paternalism in medicine and research often revolves around the special relation of physician or scientist to knowledge and the expected or actual limited capacity of subjects who lack expertise to make the same sort of informed choices that a scientist or physician can make. Procedural autonomy recognizes that diminished capacities impact the ability of agents to exercise autonomy, and so enforce certain standards of behavior to correct for it.

The notion of "informed consent," which is now central to conducting research directly involving human subjects, is based upon the idea of expanding autonomy by providing subjects with increased capacities where possible. By informing a subject about the nature of some study, their capacity to make a free choice is increased. Without proper knowledge, even explicit assent would be suspect and may not constitute free choice. We should also be mindful of the role of various pressures in creating "duress" which might negate consent and undermine true autonomy.

6.3 Dignity

Dignity is a difficult concept, inter-related in many ways with the notion of autonomy. It is perhaps best evidenced in its absence. The notion of dignity as mentioned above is tied up with the Kantian notion of autonomy. In Kant's ethics, dignity means that we deserve to be treated as "ends in ourselves" as opposed to being used by others as means to ends. In order to be consistent, we must act the same toward others. In the realm of research, this also implies that informed consent be correctly given. Because research is intended to develop knowledge that is generally useful, and may harm those who are its subjects, there is always the likelihood that a human subject in an experiment is being used as a means to some end. In fact, as part of the consent procedure, subjects must be warned that the research they are taking part in is not intended to help them in particular, and that they should have no expectation of benefit, even while they may be harmed. In such a case, a subject is always a means to some other end, and dignity may suffer. The only way to help preserve dignity is to ensure that they are fully informed about their use in the study, the likelihood of harms, and the fact that their involvement in the study is not meant and will not likely give any benefit to them.

Treating subjects with dignity, and subjects "having" dignity, are not the same, though both are considered generally required. During the course of a study, an autonomous person with dignity must have the means and knowledge to stop their involvement at any time and without duress. Subjects must be treated as though they have the capacity to consent, and treated according to their individual capacities. If capacities change during the course of the experiment, then their participation should be open to reassessment. Each of us is an individual, and even when being used as an instrumentality in some study, must be taken to be individuals with our own identities, needs, capacities, weaknesses, etc. Dignity is impaired when subjects are treated merely as data points and not as fully autonomous individuals with all the human rights to which they are due.

Dignity is necessarily involved with the next major bioethical principle: beneficence. When we treat subjects with dignity, we act with certain intentions, even if the outward appearance of our actions might be the same without recognizing dignity. We may go through the motions of an informed consent interview without ever fully recognizing that our potential subjects are fully dignified, autonomous individuals, with the sole intention of signing them up for the study. While we might have satisfied all the checklist items for an ethical study, we might well question whether we have acted ethically. While there is no institution, nor should there be, that can resolve this distinction, we should at least be aware of our inward motivations in an ongoing attempt to maintain the proper ethical relationship with our subjects, and to respect this principle of dignity.

Respecting the dignity of individuals requires that, since we are actually treating subjects as some means to an end, we nonetheless acknowledge and strive toward providing them with the best quality of life in the process consistent with our study. Care must be given to participants in order to maintain their best quality of life consistent with the study, they must be advised that they can withdraw should they experience pain or discomfort. They must be treated as though, although they are being used as some means to some end, that is not their primary nature or role, and that they continue at all times to be autonomous individuals capable of free choices.

Of course, our discussion so far relates to human subjects in specific trials, but these same principles might well be extended beyond the familiar terrain of research ethics toward subjects of all scientific studies which impact on humans in general. Autonomy and dignity may be impaired by research that is not generally considered "medical" research and thus regulated by institutions derived or otherwise related to the Helsinki Convention. In some states, any study involving humans, even non-medical, must abide by the Nuremberg/Belmont principles. Scientists wishing to acts as broadly ethically as possible might also consider the effects of their studies in non-medical sciences on individuals and populations, and ask to what extent they may be impacting on autonomy and dignity. For instance, if population data is used, can it be traced back to individuals, can it be re-identified with harm to individual dignity and autonomy? Has it been gathered consistent with respect for these values? Does a new technology threaten these values and has its testing and development required the use of participants who were properly treated with dignity and whose autonomy was respected? Underlying all of these considerations, either when directly or indirectly involving human subjects, is another question of intent: are we acting with beneficence/non-maleficence?

6.4 Beneficence/Non-maleficence

Beneficence and non-maleficence are closely related but not the same. If one acts to cause good, the value we place on that as opposed to merely not causing harm is typically different. We tend to value causing or creating some good higher than merely avoiding harms. Doing good is again part of the background expectations of science in general. We often view the net result of even the most basic research as increasing the store of human knowledge and view this as good.

Besides the general good that a researcher might feel that she contributes through the advancement of knowledge, each study that uses human subjects must be designed and intended to promote the good. Scientific investigations using human subjects must have some clear goal toward revealing knowledge or developing technologies that positively affect the public welfare. Curing diseases, relieving suffering, improving health, etc., are all positive goals and may warrant the use of human subjects in direct studies. Research for research's sake, especially where subjects may be at risk of harm, is not beneficent. It may well be neutral and non-harmful in the end, but it fails the qualification for beneficence that is necessary for a study to be ethical.

Beneficence is satisfied by both internal motivations and concrete actions and results. Research may benefit the good even if it causes harm in the process. The intent to create a new vaccine, following standard procedures, well-tested methods, based upon the state of the current art, and using animal and human subjects appropriately may yet fail and may yet injure people. This is not proof that it did not satisfy beneficence. To do so, the experiment needs to be designed with the intention to do good. Moreover, promoting the good throughout the course of the study may impact the ongoing research. If during a study there are enough signs that the study will cause harm, and that harm is not outweighed by the good of the study, then the study should be halted. Harms that occur as a byproduct of worthwhile (the good outweighs the harm) study must be remediate as possible. Participants should be treated, assisted medically or emotionally, provided for to accommodate their harms, and duly compensated for their participation. Beneficence means caring for the subjects of a study as well as for society at large, weighing the costs and benefits, and ensuring that the good outweighs the bad. Beneficence is not quite the same thing as non-maleficence, and so scholars and practitioner tend to distinguish these two values. Non-maleficence requires only that one not have any "bad" intentions so that the study ought not be intended to cause harm or promote harms. During the conduct of a study, one must also not shift intentions toward harmful ones. Harms need not be physical. Certainly harms can also be psychological or involve deprivations of rights. Harms may also affect groups rather than individuals. Studies that cause the potential or actual harms to populations may also fail for violating these principles.

Science is a public good. This means it exists due to the patience, good will, and funding of the public and ought to positively impact the public. Scientists and the public are mutually interdependent and science must somehow improve the public, whether by the acquiring of true, basic knowledge, or by increasing the health and general welfare of those who allow science to proceed. It should be conducted in light of these reciprocal duties.

6.5 Justice

Justice is another requirement of equal treatment and is linked to the notion of dignity. More than equal treatment, we are to be attended to as we are due, and we may be due differing treatment according to our unique capacities. The modern notion of justice balances principles of autonomy against the provision of certain goods. There is a tension between the good of individual liberty, which always exists, contrasted against the necessity for basic goods and unequal distribution of those goods.

Justice requires the balancing and proper apportionment among those goods in the context of ongoing conflict.

Justice is impaired both by nature and by circumstance. Unequal apportionment of capacities and opportunities for developing abilities, as well as uneven distribution of basic and secondary goods impact on the prospects of individuals. The notion of "social justice" as described by the modern legal philosopher John Rawls, involves the assigning of rights and duties and distribution of benefits and burdens in a way that is most "fair." What is considered "fair' is that which treats people with the most dignity, similarly to the notion as devised by Kant. Freedom should be maximized while at the same time we should accommodate unequal distribution of primary goods and capacities.

Consider the Tuskegee case I discussed in the first chapter. The participants of that study were not only recruited without likely understanding the nature of the study fully, nor their positions within it, but also were never apprised of the fact that they could have at some point been cured. Maximizing justice in the use of human participants requires treating those with unequal capacities in ways that maximize their freedom. The Tuskegee study subjects were generally illiterate, they were members of a class of people who were historically ill-treated and underprivileged (African Americans) and so they required a heightened degree of protection.

Justice requires that members of historically underprivileged, poorly protected, and vulnerable classes be treated with extra protection when using them as subjects in studies. When vulnerable subjects are used, they must be given special attention during the consent procedure, and where possible, using vulnerable subjects should be avoided entirely. This is not always possible, and often justice requires studies to be done that are likely to help vulnerable populations, and so using subjects from those populations is necessary and just. What is then required is special attention to their reduced or altered capacities, and careful oversight during the course of the study to ensure that along the way the harms are minimized.

Fairness recognizes that the allocations of positions and capacities is not based generally upon merit, but rather due to some natural lottery. We are born into our positions in society without any choice of whom we might be, what natural and social advantages we will have, etc. Without diminishing the role and justice of rewards for merit, we can and should recognize that people are treated unfairly through no fault of their own, with access to differing positions and goods having little to do in the first instance with merit, hard work, virtues, or other qualities within our controls. The role of a just society is to help even out the playing field, making up for the natural lottery, and opening positions and opportunities to as many as possible, as well as recognizing and rewarding merit when it is obtained without infringing justice.

Justice can be infringed even when people are treated (objectively) equally given that the principle recognizes that we begin from unequal positions not as a matter of choice or desert. Thus justice may require special treatment for disadvantaged populations and individuals. As in the case of the Tuskegee study, the subjects were particularly vulnerable given their capacities and social statuses. Merely giving informed consent in the same manner as in other studies would have been insufficient given their educational backgrounds. Special care beyond what might have been necessitated in the same study using individuals with better educations would be necessary to properly protect the dignity and autonomy of the study participants. The same would be true if children or other vulnerable populations are used. The level of care and attention cannot simply be equal to be just, rather it must be suited to the participants.

6.6 The Four Principles and "Care" Ethics

Beyond the four principles typically associated with bioethics are numerous other approaches to ethics, the good, and the treatment of others in the course of medicine and research. One specifically worth noting when we touch on issues of bioethics is "care ethics." This approach is often contrasted with the main theories we have discussed in this book. Unlike those theories, the methodological assumptions about the nature of the good differ in care ethics from standard, normative theories. Specifically, the good is not an abstract that we can derive from some first principles in the manner of other logical and material truths sought by philosophers. Rather, the good is contextual. In the world of medicine and biomedical research, for instance, the duty of care flows from the nature of the relationships among physicians and patients. Care in that context is defined as a process, value, or virtue and not as some deductive truth. It is something we must attend to as we go about any activity linked to health. Where science involves human health, the duty of care also follows. The duty of care binds us to act in ways that support the well-being of others. It requires attentiveness to the disposition and comfort of others, especially those in need, it imposes a responsibility and willingness to respond to needs of others, it demands our competence in delivering goods to those in need, and we must be responsive and not passive.

Care is an emotional state or readiness on top of attentiveness, and care ethics views those in positions that are especially able to be responsive as best engaging their ethical duties when they proactively provide care for others over whom they are best suited to care. This is related to the notion of paternalism. Both care ethics and paternalism are often criticized for being insufficiently focused on the primacy of liberty.

Science, and especially medical science, run the risk of "paternalism" due to the specialized nature of the expertise and access to knowledge enjoyed by their practitioners. As well, scientists and physicians are often viewed by others as having not only special knowledge but authority by virtue of their positions. Paternalism follows from the use of authority as the basis for obedience. If liberty is a primary good, then we ought to be suspicious of all recourse to authority. Moreover, we should be suspicious of modes of behavior that demand we give up our freedom to act, such as with care ethics which demands we act even when it is not in our personal interest.

Are researchers free to act in their own interests, or must they be forced to act always in the interests of others, even when it contravenes their own freedom? This is a basic debate in ethics, and especially in political theory. The debate has shaped the discourse of ethics for the past fifty years and underlies much of the discussion of the role and duty of science in society. Meanwhile, institutions demand that we develop norms of behavior, that society and scientists have some predictability in their interactions, and that progress not be hindered while we debate. Rather than being relegated to the halls of the academy, the ongoing discussion about the duties and limits of science and scientists is being worked out through various institutions and their rules, regulations, and laws. It is also the focus of education. Increasingly, scientists are being introduced to concepts, cases, and discussions like those we have touched on only briefly in this book for the sake of empowering them to sort out some of the ethical issues underlying their daily practices, rather than simply setting down rules and methods of obedience and punishment.

By providing examples, principles, and discussing cases, we form the basis for a more educated and thoughtful recognition and discussion of issues as they arise. Ethics is not abstract anymore. History is filled with examples of science going wrong, people being injured and their rights abridged, and the public losing confidence of those they give the power and privileged of examining nature's secrets. The risks go well beyond injury and death of human subjects. Humanity itself is the subject of our ongoing examination of nature and her laws. We all benefit or are harmed through the practices of science, and we must be aware and careful about its repercussions. Even where we do not personally care about others, or especially those who seem remote from our studies, we should at the very least act out of enlightened self-interest and protect the boundaries of science itself, cognizant of the interdependence of its institutions with the general public, and mindful of its vulnerability at the hands of a public that loses confidence in it with each new avoidable tragedy or public fraud.

In the next chapter I will investigate in more detail some of the institutions that have been built to help guard against error and harm, and that now police the borders of ethical conduct where individuals fail.

Study and Discussion Questions

1. How do the Nuremberg/Belmont principles help in sciences beyond medical science? What sciences (if any) are immune from ethical considerations and why?
2. Describe how autonomy and dignity are related principles, and what ethical theories are involved in their understanding and application.
3. How does justice in ethics, beyond human subjects research, alter the manner in which research is undertaken and applied through technologies?
4. Consider and describe difference between paternalism, autonomy, and care ethics.

References

Allmark, Peter. 1995. Can there be an ethics of care? *Journal of Medical Ethics* 21(1): 19–24.
Andorno, Roberto. 2009. Human dignity and human rights as a common ground for a global bioethics. *Journal of Medicine and Philosophy* 34(3): 223–240.
Ashcroft, Richard E. 2005. Making sense of dignity. *Journal of Medical Ethics* 31(11): 679–682.
Benatar, Soloman R. 2001. Justice and medical research: A global perspective. *Bioethics* 15(4): 333–340.
Engelhardt, H. Tristram. 1996. The foundations of bioethics. Oxford: Oxford University Press.
Foster, Charles. 2013. Human dignity in bioethics and law. *Journal of Medical Ethics*: medethics-2013.
Hamington, Maurice. 2004. *Embodied care: Jane Addams, Maurice Merleau-Ponty, and feminist ethics*. Urbana: University of Illinois Press.
Harris, John. 1999. Justice and equal opportunities in health care. *Bioethics* 13(5): 392–404.
Häyry, Matti, and Tuija Takala. 2005. Human dignity, bioethics and human rights. *Developing World Bioethics* 5(3): 225–233.
Kaczor, Christopher. 2006. *The edge of life: Human dignity and contemporary bioethics*, vol. 85. Dordrecht: Springer.
Kass, Nancy E. 2004. Public health ethics from foundations and frameworks to justice and global public health. *The Journal of Law, Medicine & Ethics* 32(2): 232–242.
Koepsell, David. 2010. On genies and bottles: Scientists' moral responsibility and dangerous technology R&D. *Science and Engineering Ethics* 16(1): 119–133.
Loewy, Erich H. 1989. Beneficence in Trust. *Hastings Center Report* 19(1): 42–43.
Macklin, Ruth. 2003. Dignity is a useless concept: It means no more than respect for persons or their autonomy. *BMJ: British Medical Journal* 327(7429): 1419.
Millum, Joseph, and Ezekiel J. Emanuel. 2012. Global justice and bioethics. Oxford: Oxford University Press.
O'neill, Onora. 2002. *Autonomy and trust in bioethics*. Cambridge: Cambridge University Press.
Rancich, Ana María, et al. 2005. Beneficence, justice, and lifelong learning expressed in medical oaths. *Journal of Continuing Education in the Health Professions* 25(3): 211–220.
Rennie, Stuart, and Bavon Mupenda. 2008. Living apart together: Reflections on bioethics, global inequality and social justice. *Philosophy, Ethics, and Humanities in Medicine* 3(1): 25.
Secker, Barbara. 1999. The appearance of Kant's deontology in contemporary Kantianism: Concepts of patient autonomy in bioethics. *Journal of Medicine and Philosophy* 24(1): 43–66.
Silvers, Anita, David T. Wasserman, and Mary Briody Mahowald. 1998. *Disability, difference, discrimination: Perspectives on justice in bioethics and public policy*, vol. 94. Lanham: Rowman & Littlefield.
Taylor, James Stacey. 2010. *Practical autonomy and bioethics*. New York: Routledge.
Wolpe, Paul Root. 1998. The triumph of autonomy in American bioethics: A sociological view. *Bioethics and Society: Constructing the Ethical Enterprise*: 38–59.

Chapter 7
Ethics Committees: Procedures and Duties

Abstract Ethics Committees (also called Institutional Review Boards, or IRBs) are common fixtures with which researchers around the world are generally quite familiar. In much of the scientific world, they are mandated by laws and regulations and research involving human subjects is generally guided and overseen by such committees. The manners in which they operate vary around the world, although the basic standards they apply generally adhere to the Belmont principles described above. In this chapter, I provide an overview of good study design, ethical use of human and animal subjects in clinical trials, and discuss briefly the principles behind informed consent. As with other chapters, I also provide guidance and context according to the Mertonian norms, extending the argument that these may form the basis for ethics in science very generally.

7.1 Legal and Regulatory Framework

The most prominent form of institution developed as a result of the ethical lapses we have only briefly begun to touch on in this book is the Ethics Committee. Guided by both the principles enunciated in the Nuremberg Code and the Declaration of Helsinki, ethics committees throughout the world have been created and act in order to help prevent the sorts of harms that injure both the public and the public's perception and trust in science.

In the wake of Nuremberg, and continued failures by scientists over time to abide by the duties expressed by the court at Nuremberg, both national and international institutions began to try to formulate local rules and codes, as well as to form institutions capable of policing and enforcing them. The Declaration of Helsinki formalized some of the principles enunciated at Nuremberg, and has become an internationally guiding document for the creation of institutions intended to protect human subjects and extend bioethical principles. The declaration is not, however, legally binding upon its signatories. It is rather meant as guidance for local laws and rules. Nations are free to institutionalize the principles enunciated in manners of their choosing, although signing onto the declaration evidences some support for the principles.

There are now a number of differing legal and regulatory frameworks under which human subject experimentation is regulated, both internationally and nationally. One is the International Conference on Harmonization of Technical Requirements for Registration of Pharmaceuticals for Human Use. Because international sales of pharmaceuticals requires compliance with these requirements, locally and nationally ascribing to certain procedures that guarantee the ethical use of human subjects in clinical trials is necessary. Other, similar frameworks exist, and in order for them to conform to the Helsinki Declaration, national codes establish ethics committees to review, consider, comment on, and guide protocols and trials. Because the international framework remains unbinding legally, nations are free to form their own manners of enforcement of the principles discussed. Nonetheless, certain manners of operation appear to be rather general and widespread, including: consistency with the Helsinki Principles, benefits ought to outweigh the risks, rights and safety of subjects must be protected, human trials should be based upon animal trials, protocols should be well described and scientifically sound, conflicts of interest should be declared, medical care of subjects is primary to the trial and its results, all participants in research should be properly trained and qualified, data should be stored to protect privacy, confidentiality should be maintained, consent must be free and informed, etc.

Ethics committees are generally composed of a range of experts as well as community members, and they serve without pay (typically) and have no stake in the outcome of the protocols they review. Their independence and objectivity must be paramount, and they should have some background in both the sciences involved and the bioethical principles to be applied. Members ought to undergo some ongoing training, including by keeping up to date on cases as they are reported, and engaging in some discussion of emerging issues in bioethics, human subjects research, or animal use in experiments if they happen to sit on an animal research ethics committee.

These are all very general considerations. Below we will look in more depth at some best practices in ethics committee procedures and considerations, as well as compare the practicalities with the principles we have discussed above.

7.2 Some Best Practices: Ethics Committees in Biomedicine

Clinical research has now been divided into various phases, each requiring differing levels of oversight, and often undertaken by different types of institutions. The expertise and level of scrutiny afforded trials changes as the trial proceeds. The first phase of all biomedical research is "pre-clinical." At this phase, there are yet no human or animal subjects and the study is considered to be focused on the very basic science that might later be involved in the development of some drugs or treatment. When some potential intervention, drug, surgery, or other treatment arises from such basic science, then a clinical trial may be devised. In an evidence-based medical scientific environment, all clinical studies must first be validated by some basic study that suggests the need or potential benefit from pre-clinical study. Once validated by sound scientific evidence that a potential benefit may arise through

7.2 Some Best Practices: Ethics Committees in Biomedicine

clinical study, a study must be approved and testing on animal subjects must be initiated.

Animal studies tend to be regulated by national laws, and ethics committees routinely review animal studies to ensure that they are conducted properly depending on the local laws and regulations. A large amount of clinical testing occurs for pharmaceuticals, and much of the regulatory and ethical oversight involved in clinical trials involves ensuring that potential drugs are tested properly. Animal testing must first be done using suitable animal models, and minimizing the harms to the test subjects. While much of the basic science in drug discovery is done in academic institutions, most of the current development (where potentially therapeutic compounds are turned into marketable drugs) is conducted by industry in conjunction with clinical centers, hospitals, and research universities with medical schools. The progress through various phases of testing for potentially beneficial compounds and devices is difficult and expensive, and most drugs do not make it to market as a result. Human clinical trials are divided roughly into pharmacology studies, exploratory, and confirmatory phases, matching roughly the moves from pre-clinical to human testing.

Most compounds and devices that do get approved for human use endure about 20 different trials before doing so. Human risk and safety is the primary concern of most trials, and what might constitute overkill for a scientific study is now standard operating procedure for interventions that are meant for the human marketplace. At every phase, good scientific practice must be ensured, but also the unnecessary harm and suffering to humans, both in the trial and intended as eventual consumers, must be closely monitored and avoided consistent with the Nuremberg Principles. This does not mean that no harm may come to subjects or even eventual consumers. As we have seen, weighing risks versus benefits is part of the ethical duty of scientists engaged in their research.

Measuring and avoiding risks both potential and actual consumes a large amount of the energy of those designing clinical studies and policing them. Not all trials are equally risky, and when a trial is deemed to pose a "minimal risk" the degree of ethical oversight changes. The potential harm of a drug or intervention can be measured by looking at: the cumulative clinical experiences involved with a drug or related drug, the results of targeted testing on a small sample or animal model, and the general known biological characteristics of the article tested. The initial stages of testing of an article, especially if unrelated to previously tested articles, are the most risky, and so the level of scientific and ethical scrutiny is particularly intense at these stages.

Typically, when an article is first tested on humans (following appropriate animal model studies) the first test subjects are healthy volunteers rather than patient/targets. After this and assuming that the article has not proven unduly hazardous, exploratory studies are then begun on targeted subjects of the population eventually intended to be the consumers. If early studies show that the article is ineffective or unduly hazardous, then further studies are typically terminated. The degree of risk that might be tolerated varies according to the degree of concern over the intended target population. That degree of concern shifts generally according to concern about the harms from the disease or other malady targeted. Terminal cancers may warrant more risk in a study than headaches, for example. As well, differing target populations and test subjects may alter the calculation and acceptability of risks. Children, the mentally ill or otherwise vulnerable may be at greater risk than others, not just due to physical or chemical vulnerabilities, but also due to institutional or other social causes.

Once pharmacological and toxicity studies have been done, a targeted population may be enlisted for a clinical trial, but only once a protocol has been drafted and submitted for ethical review. The protocol must meet the criteria first of scientific acceptability. It must be designed according to the best available data, methods, and with appropriate empirically-justified controls and procedures, as well as fail-safes to ensure that the results of the study are of sufficient scientific value once concluded. Details about the protocol must be scrutinized for their scientific merit even as the study is checked for ethical sufficiency. A study protocol that lacks scientific merit cannot be approved. A scientifically worthy study then may be classified as either "minimal risk" warranting little more in the way of ethical scrutiny, or more than minimal risk warranting a full review by an ethics committee. Before a study is classified as either minimal risk or more, types of risks must be classified and assessed. Not all risks, even significant risks, are bodily or "physical." Some unacceptable risks may include psychological or even social and economic harms. If taking all risks into consideration the probability and magnitude of anticipated risks is no greater than those that potential subjects are exposed to in their day-to-day lives under routine circumstances, then typically the review of a study is expedited, not requiring a full committee review. For greater than minimal risks, a more thorough procedure is employed.

7.3 Minimizing Risks: Stakeholders and Duties

Typically, there are numerous stakeholders involved in the development of new medical articles, including scientists, government agencies, drug companies, research institutes, funding agencies, patients and patient rights groups, among others. All of these parties tend to have some interest in the development and marketing of an article, and their participation in the positive and negative responsibility associated with their roles may need to be considered at various stages of discovery and development. Governmental and other regulatory agencies may be responsible for reviewing and approving protocols as well as monitoring their activities for compliance. Funders and other sponsors may be responsible for obtaining proper consent, protecting the rights of subjects and ensuring the responsibility of investigators, ensuring ethics committee review, maintaining transparency with the ethics committee, being aware of and disclosing potential conflicts of interest, etc. Investigators need to abide by protocols, ensure that their protocols conform to scientific and ethical standards, maintain transparency with funders and other sponsors as well as governmental and other regulatory bodies, be versed in the state of the art of their science, communicate with subjects in ways that they can understand about the science and their roles in the study, as well as other ongoing duties.

While the roles of each of the various stakeholders all involve ethical responsibilities at various points, the role of the ethics committee and that of the investigators are most closely associated with ethical responsibilities throughout a trial. Ethics committees must in general oversee a study, taking responsibility for potential subjects and ensuring as the study proceeds that not only has it been commenced under proper conditions, but that it proceeds through its lifespan as ethically as possible. Among the duties of ethics committees and their members are: protecting the rights of subjects paying special attention to potentially vulnerable subjects, protocol review and criticism both for scientific and ethical validity, analysis of proposed investigators and their abilities in respect of the proposed study, ongoing review and oversight during the course of a study paying attention also to changes in the regulatory and ethical milieu, paying attention to adverse event reports and being willing to shut down a study, calling in researchers for further questioning, and other appropriate measures to protect subjects from harms both physical and psychological.

Ethics committee meetings should adopt standardized procedures conforming with best practices as currently understood by peers, should be conducted professionally by members who attend having read the protocols in depth before their meetings and with questions. All members should participate, led by a chair who understands well the role and importance of a thorough review as well as the principles involved in a proper review. The committee should be composed of peers from throughout the institution, versed in various appropriate sciences as well as representing a background in ethical issues.

The participants or subjects are another major set of stakeholders, and while many duties are owed to them, they also have various explicit duties regarding their roles. They have a duty of transparency when enlisting in a study. Part of the process of enrolling subjects involves interviews and questionnaires, and they must be candid in their responses. Their safety and the proper conduct of the science are at stake and those managing the study cannot ensure either if subjects are not transparent.

Recently, a new type of stakeholder has emerged called a Clinical Trial Services provider, or Clinical Research Organization. These are private companies that can be hired to do outsourced parts of trials or manage the trials themselves. These organizations run and manage the trial, at either the animal or human stages, and take on the responsibilities of any other organization or institute that would be in their place by recruiting subjects, overseeing a trial's conduct, as well as managing risks and reducing and reporting adverse events. There may be a myriad of other involved parties working with those who are managing or conducting trials, including centralized laboratory services, study sponsors and their partners, etc. All of these relations and interactions should be carefully monitored for potential involvement with human subjects and their rights, as well as for the possibility of lapses. One complicated area of potential harm regards data, and another party may be involved to monitor and protect data throughout a study: a Data Safety and Monitoring Committee.

An independent DSMC can provide comment and oversight for complicated data management issues that arise, now especially likely in light of multi-center, multi-party studies involving collection of significant amounts of data. Such a committee can work in conjunction with an ethics committee, and may be employed *ad hoc* depending upon the nature of the protocol and whether it warrants it.

All of these various stakeholders have ethical duties, perhaps at differing stages, throughout the course of a clinical trial. Careful understanding of the nature of the relationships of each to the other, and especially to subjects involved in studies as participants, can help to avoid ethical errors and resulting harms. As well, understanding the nature of and importance of clinical trials in the conduct of applied research can help to put into context the various duties owed.

7.4 Clinical Trials: Methods and Duties

A good amount of the work in determining the ethics of a clinical trial involves careful study of the trial design and protocols. There are scientifically sound ways to conduct trials, and an investigator's primary duty to science is to ensure that a study is conducted properly. Ethics often follows proper study design. Scientific studies cannot prove hypotheses, they can only provide confirming support for them or falsifying data. Understanding the role of a clinical trial depends initially on understanding its role in the institutions of science. Failure to understand this may involve making significant errors in design and interpretation, and failures to properly report the results of a study to both a community of peers and to the public. Sound clinical trial design requires proper use of control groups, randomization, and blinding all to help remove possibilities for researcher error. Failure to use these three "cornerstones" of good clinical trial methodology may result in ethical lapses as well. Ethics committee members should concentrate on these three aspects of good clinical trial design in reviewing protocols, although their inquiry does not end there.

Clinical trials are started when, having done basic research into molecules, devices, and methods and sufficient safety testing in pre-clinical studies, researchers have reason to move forward to testing the efficacy of an article on human subjects. Their primary objective is to determine the safety and efficacy of an article meant for use in human medicine. One of the most important considerations in trial design is proper understanding of the trial outcome/endpoint, and accounting for this in the trial design and protocols. It is crucial to understand the role of statistics in both confirmation and falsification, and the trial design must therefore use a proper sample size, as well as subjects who are good models considering the nature of the article being tested. Two major types of errors can result from improper design, including: (1) false positive results, and (2) false negative results, and selection of appropriate samples and subjects can result in either. Thus an initial duty of investigators and ethics committee members is to ensure that the study design is optimum for biostatistical purposes. Aside from these two errors, two proper conclusions may come from proper trial design and conduct, conclusions that more properly reflect nature and that will improve the progress of medicine: that the proposed and tested treatment is effective, or that it is ineffective. The trick of ethical study design is maximizing the chances of the latter two types of results while minimizing the risks to human subjects in the process.

To help ensure this situation of maximal chances of helpful results and minimal risks to subjects in the process, an essential standpoint of clinical equipoise must be maintained by investigators and ethics committees, as well as other stakeholders. This means that in order for the trial to work correctly, there should be substantial uncertainty about which "arm" of a blinded and randomized trial with controls a particular subject will be in. Investigator equipoise serves a similar role, helping to ensure that investigators are not inclined one way or another to suspect that a particular test article will prove beneficial or not. Equipoise essentially means that we remain non-committed throughout the trial as to which of the two beneficial results will follow from the trial, and accept that negative results (e.g. an article is ineffective) are valid and worthy scientific results from any study.

Using controls appropriately invokes ethical duties. Control groups should be chosen from the same population as test groups, and should be similar in ways that can scientifically help to verify or falsify the proposed treatment article. Four types of controls can be created: placebo, no treatment, different dose or regimen, or standard treatment. In a placebo trial subjects are randomly assigned to either a test treatment or identical-appearing non-test treatment (receiving a placebo, without knowledge) most such trials are double-blinded so that neither the subject nor the researchers know which subject is receiving the placebo during the course of the treatment. In a no-treatment trial, subjects are randomly assigned to test treatment or no treatment at all. This sort of trial is not blinded and is only suitable when it is too difficult or impossible to blind a study. In a randomized, fixed-dose, dose-response trial, subjects are randomized into groups receiving differing doses. Such trials are double-blinded. In an active control trial, subjects are randomized into either test treatment or active treatment of a kind ordinarily administered to those with their condition. The trial is generally double-blinded and meant to demonstrate whether the test is as good or better than standard treatment. A trial may use any number of types of control groups appropriate for achieving the scientific results during the course of the trial.

The Declaration of Helsinki states: "The benefits, risks, burdens and effectiveness of a new intervention must be tested against those of the best current proven intervention, except in the following circumstances: the use of placebo, or no treatment, is acceptable in studies where no current proven intervention exists; or where for compelling and scientifically sound methodological reasons the use of placebo is necessary to determine the efficacy or safety of an intervention and the participants who receive placebo or no treatment will not be subject to any risk of serious or irreversible harm. Extreme care must be taken to avoid abuse of this option." Thus, there is generally no ethical issue with using a placebo control group if the condition for which the article being tested has no effective treatment. Using a placebo control group may pose ethical issues if there is an available treatment that is both safe and effective. An exception may be if the only known treatment is risky and where the risks of such treatments justify exploring a potential treatment article that poses fewer risks according to pre-clinical studies. In both placebo and no-treatment trials, subjects are to receive medical monitoring and treatment. Placebo-controlled trials measure the total mediated effect of treatment while active control trials measure the effect relative to another treatment. They also make it possible to distinguish between adverse events caused by both the drug and underlying disease. Ethics committee members should be aware of special issues involved with studies involving placebos and non-treatments. In the case of differing dose vs. standard treatment controls, subjects will be randomized to either test article or standard treatment groups, and so will not be denied some treatment. Moreover, they are going to be medically monitored and treated throughout the study.

7.5 Randomizing and Blinding

Randomization helps to avoid selection bias which is the result of preferred enrolment of certain subjects over others. By basing the assignment to groups based upon chance, errors based on conscious or unconscious preference for certain subjects may be better avoided. The chances of assignment should be as close to 50 percent as possible and is now typically conducted by computerized means before the trial is commenced. As well, it should be conducted by a third party, someone not involved in the trial. The randomization list should be kept by someone not involved in the trial to help ensure that the stakeholders involved in conducting the trial remain unaware of which subjects are in which group. The only parties with access to the randomization list should be the investigators or technicians involved in preparing the trial drugs and the DSMC (in case of adverse events). A copy of the treatment code should be available at all times just in case it becomes necessary to break it for a participant, either by unblinding a sealed envelope or through a telephone based unblinding process.

Randomization can be done by apportioning unequal numbers of participants to differing treatment groups so that similar characteristics of importance to the study are present for all treatment groups. Stratified randomization is used to help ensure that the numbers of males and females are roughly equal in both groups, or that the numbers of subjects at similar disease stages are present.

Blinding helps to ensure equipoise by keeping knowledge about the makeup of the study groups and those receiving which articles (the test article vs. a placebo or standard treatment) from unintentionally affecting the interpretation of results. Any factor that may allow for deciphering whom is receiving which article should be carefully hidden as well. In one famous case, a technician was able to determine the distinction between test articles and placebos, by shaking a vial, and then altered the protocol by choosing which article to administer. This trial was both ineffectively blinded and unethically altered by an investigator or technician.

There are different levels of blinding: Single blind usually means one of the three categories of individuals remains unaware of intervention assignments throughout the trial. Double-blind means that participants, investigators and assessors usually all remain unaware of the intervention assignments throughout the trial. In medical research, however, an investigator frequently also makes assessments, so in this instance, the terminology accurately refers to two categories. Triple blind usually means a double-blind trial that also maintains a blind data analysis. Knowledge of treatment may have an influence on: Recruitment of participants, treatment group allocation of participants, participant care, attitudes of participants to the treatment, assessment of endpoints, handling of withdrawals, exclusion of data from analysis, and statistical analysis. The proper conduct of science demands that both randomization and blinding be used appropriately, with proper ethical controls, so that outcomes can be most adequately measured. Proper use of these tools can help to avoid investigator bias, evaluator bias, and performance bias, all of which can introduce errors into a study. Investigator bias occurs when an investigator inadvertently or intentionally favors one group at the expense of another. Evaluator bias occurs when interpreting the data where an evaluator favors one set of data over another. Performance bias occurs when a participant knows which group he or she is in and may skew measurements or leave the study because of their knowledge.

7.6 Some Ethical Issues in Clinical Trials: Risk vs. Benefit

As discussed above, ethics committees must delve into the science involved in a proposed trial and make some judgments about its viability, risks, and likelihood of generating knew knowledge. They must also weigh risks and benefits. In sum, the ethics committee should focus on the science, ethics, and quality assurance of a proposed study. Among the many questions the ethics committee must delve into are the following:

Has the study been reviewed or approved (or declined) before? Are the investigators properly qualified to do the study? What is the known safety of the test article? What is the scientific rationale behind the study? What are the expected benefits of the test article in normal clinical care? Can the use of a placebo be justified in the study and how? Is this an exploratory or confirmatory study? Have the best sample groups been identified and can they be recruited? Does the study appropriately randomize and blind? Has the proper sample size been calculated and how? What assumptions justify the proposed sample size? Will there be enough participants? Can the trial be completed with the available resources?

Besides these, the committee must do a risk/benefit calculation of sorts. "In medical research involving human subjects, the well-being of the individual research subject must take precedence over all other interests." "Medical research involving human subjects may only be conducted if the importance of the objective outweighs the inherent risks and burdens to the research subjects." (Declaration of Helsinki). The benefits calculated, while taking into account the potential and the costs of harms to subjects, are calculated rather from the point of view of society in general. Will the study advance science and improve health and quality of life? All known instances of use and testing of an article should be referenced in the protocol so that the committee can do its risk calculation effectively, although the committee may have a positive duty to do a bit of research in this area as well. The degree of review and scrutiny may increase depending upon the riskiness of the use and testing of the article proposed. Risks may involve physical, mental, social, societal and economic harms among others, and should all be calculated as part of the committee's assessment. There is no set objective measure for the calculation of risks given the various dimensions and types of harms that may occur due to the use of any article on any number of subjects. As well, potential benefits should be similarly calculated and can also include non-physical, non-medical benefits.

Risks naturally fall as the study of an article moves among its phases, from phase I to Phase III, given that during the course of such study the potential harms become less mysterious and data is generated. Still, at every phase, ethics committees have the same duty to avoid risks and harms, to weigh their potentials against potential benefits, and to avoid unnecessary harms while maximizing societal and scientific benefit.

7.7 Informed Consent

Of primary importance in the design and presenting of a protocol to an ethics committee is the assurance that study participants will be given proper information and that their consent will be freely obtained. Informed consent involves more than generating a checklist of items to be disclosed. It also involves ongoing careful interaction with participants regarding the study and their roles in it. In layman's language, the following issues at least should be addressed with the potential subjects: that the trial involves research; the purpose of the trial; trial treatment and procedures; the participant's responsibilities; experimental trial aspects; foreseeable risks or inconveniences; expected benefits; alternative procedure(s) or treatment(s); compensation or treatment available in the event of trial-related injury; payment to participant; expenses for participant; that participation is voluntary and the participant may refuse to participate or withdraw from the trial at any time; that the monitor(s), auditor(s), the ethics committee, and the regulatory authority(ies) will be granted direct access to the participant's medical records; that records identifying the participant will be kept confidential; that the participant or representative will be informed if information becomes available that may be relevant to their willingness to continue participating in the trial; names of person(s) to contact for further information regarding the trial, rights of trial participants, and in the event of trial-related injury; the circumstances or reasons under which participation in the trial may be terminated; expected duration of trial participation; and approximate number of participants involved in the trial.

In getting informed consent, it must be clear and safeguards should be incorporated that participants are not only reading and signing the forms properly, but that they *understand* fully what is going on. This involves some interviewing and discussion in which participants are able to ask questions as well, including about the reasons for the trial, their own benefits, the risks faced, the length of the trial, the discomfort they may face, other treatment options, and their rights to exit the trial at any time.

7.8 Vulnerability and Justice

There is an inherent conflict in the desire to provide better treatments to vulnerable populations and the need to use vulnerable subjects in studies. Vulnerability often implies diminished capacities and so the ethics committee must take special care in reviewing and conforming a proposed study and its consent form to account for potential vulnerabilities. Children, the mentally handicapped, underprivileged and elderly and others with diminished capacities as well as potential sources of duress flowing from these states must all be carefully scrutinized and where possible accounted for so that vulnerable individuals and populations are justly treated in the course of a study.

Study and Discussion Questions

1. Which types of biases are blinding and randomization meant to address? What is the role of the ethics committee in reviewing blinding and randomization in a proposed protocol?
2. What are the two important errors that can result from improper study design? Is improper design an ethical issue? Why or why not?
3. What are the differences between minimal risk studies and more than minimal risk, and why is there a different level of ethical scrutiny for the two types?
4. What sorts of actions must researchers take to help ensure informed consent? Why are vulnerable populations treated differently for gathering consent, and how can we best ensure that they are treated fairly?

References

Angell, Marcia. 1997. The ethics of clinical research in the Third World. *New England Journal of Medicine* 337: 847–848.

Corrigan, Oonagh. 2003. Empty ethics: The problem with informed consent. *Sociology of Health & Illness* 25(7): 768–792.

Emanuel, Ezekiel J., David Wendler, and Christine Grady. 2000. What makes clinical research ethical? *JAMA* 283(20): 2701–2711.

Emanuel, Ezekiel J., et al. 2004. What makes clinical research in developing countries ethical? The benchmarks of ethical research. *Journal of Infectious Diseases* 189(5): 930–937.

Goodyear-Smith, Felicity, et al. 2002. International variation in ethics committee requirements: Comparisons across five Westernised nations. *BMC Medical Ethics* 3(1): 2.

Hellman, Samuel, and Deborah S. Hellman. 1991. Of mice but not men: problems of the randomized clinical trial. *New England Journal of Medicine* 324(22): 1585–1589.

Jonsen, Albert R., Mark Siegler, and William J. Winslade. 1982. *Clinical ethics: A practical approach to ethical decisions in clinical medicine*. New York: McGraw-Hill.

Keith-Spiegel, Patricia, Gerald P. Koocher, and Barbara Tabachnick. 2006. What scientists want from their research ethics committee. *Journal of Empirical Research on Human Research Ethics* 1(1): 67–81.

Manson, Neil C., and Onora O'Neill. 2007. *Rethinking informed consent in bioethics*. Cambridge: Cambridge University Press.

McGee, Glenn, et al. 2002. Successes and failures of hospital ethics committees: A national survey of ethics committee chairs. *Cambridge Quarterly of Healthcare Ethics* 11(01): 87–93.

Miller, Franklin G., and Howard Brody. 2003. A critique of clinical equipoise: Therapeutic misconception in the ethics of clinical trials. *Hastings Center Report* 33(3): 19–28.

O'Neill, Onora. 2003. Some limits of informed consent. *Journal of Medical Ethics* (29)1: 4–7.

Peto, Richard, et al. 1976. Design and analysis of randomized clinical trials requiring prolonged observation of each patient. I. Introduction and design. *British Journal of Cancer* 34(6): 585.

Ramcharan, Paul, and John R. Cutcliffe. 2001. Judging the ethics of qualitative research: Considering the 'ethics as process' model. *Health & Social Care in the Community* 9(6): 358–366.

Rosenberger, William F., and John M. Lachin. 2004. *Randomization in clinical trials: Theory and practice*. New York: Wiley.

Valdez-Martinez, Edith, et al. 2006. Descriptive ethics: a qualitative study of local research ethics committees in Mexico. *Developing World Bioethics* 6(2): 95–105.

Verástegui, Emma L. 2006. Consenting of the vulnerable: The informed consent procedure in advanced cancer patients in Mexico. *BMC Medical Ethics* 7(1): 13.

Yank, Veronica, and Drummond Rennie. 2002. Reporting of informed consent and ethics committee approval in clinical trials. *JAMA* 287(21): 2835–2838.

Chapter 8
Duties of Science to Society (and Vice Versa)

Abstract Science is an amorphous, distributed, and dynamic institution, composed of many other institutions and falling under the control of no central body. Rather, the body of knowledge that science develops becomes a part of our common heritage. Over time, as science improves our understanding of the universe and our place in it, we are enriched in ways that are both tangible and intangible. Because of its nature as an institution composed of institutions, with many connections both tangential and integral to nearly every part of society, we must be particularly mindful of the value, impact, and responsibilities of science and those working in it. As well, we should take care to relate the reciprocal duties of science to society and vice versa. Scientists do not work in a vacuum, and the work that scientists do benefits us all, whether we know it or not. It is incumbent upon scientists to communicate with the public, and to interact in ways that are both educational and ethical because science and the public stand in mutually beneficial relationships to one another, and are also mutually dependent. In this chapter, I consider to what extent science and society owe duties to each other.

8.1 Science and Society

As we have seen in the examples we have discussed, cases and histories of various sciences, and in recent history, sometimes scientists behave in ways that bring disrepute upon them and their professions. When they do so, confidence in the sciences inevitably diminishes, often deservedly. Practicing our professions, investigating nature and society, depends upon a trust placed by the general public in scientists who will pursue the truth dispassionately, and with an eye toward the general good. There is no right for science to subsist on the weal of the public, and it is a great honor to be entrusted with the ability to do so as a scientist – to delve into nature and her mysteries not because there is potentially some monetary profit or material good to be obtained, but because we care about the search for truth as a good in itself.

I have framed the discussion of ethics in scientific research and scientific integrity in terms of the Mertonian norms of science, although there may well be some debate about those norms within both the scientific and philosophical communities. However, Merton's attempt to define those norms is primarily descriptive and not normative. That is to say, he attempted to describe *how science actually works*, when it works best, rather to define a code of behavior from first principles. The values of universalism, communalism, disinterestedness, and organized skepticism describe stances and practices that help move science forward. Opposing these values has historically held science back. I expect that scientists and the public will jointly agree that the steady and successful progress of the search for truth works to our mutual benefit, and that both the simple fact of knowing better about our universe, and the practical consequences that sometimes flow from that knowledge, mean we have a shared interest in science and its institutions. We also care about society, which is the context in which science operates. Embracing the Mertonian norms, I have argued, means embracing certain moral values too. Even if we do not care about ethics *per se*, our mutual interests in science demand certain behaviors that we often associate with ethics or morality.

But what do scientists owe society, besides conforming to certain behaviors according to the Mertonian norms? How and why should science regard society in certain ways, and according to what principles? Are there positive duties to engage with society in certain ways as well as negative duties to avoid certain behaviors? If so, what are the parameters for those duties? Working as we have from the standpoint of the Mertonian norms, I will look below at some ways in which those norms suggest certain types of behaviors and stances necessary in the relationship between science and the public.

8.2 Universalism

In science, universalism means that science is the same wherever we go. There are no local truths, but rather a natural world that can only be understood by viewing its laws as universal. There are truths that can be discovered regardless of our particular beliefs, and it is the project of science to uncover them regardless of our local prejudices. The objects of science are unaffected by culture or history. The institutions of science should thus be engaged in the same work everywhere, and accept as a matter of course that wherever we go in the world, other scientists are pursuing the same underlying truths in similar manners, and that the results of these studies will lead us all, jointly or severally, to the same basic facts about the universe. Failures to abide by this basic notion can be disastrous. A stark example of this is Lysenkoism, the failed Soviet science based upon Lamarckianism that hued most closely to communist orthodoxy, but the pursuit of which led to the deaths of perhaps hundreds of thousands due to starvation in Stalinist USSR.

The Darwinian theory of natural selection did not conform to communist notions of perfectibility and cooperation. The notion that populations grow stronger through competition, and the weeding out slowly of the weaker by natural forces, competition for scarce resources, and the slow halting progress of evolution, runs counter to many of the precepts of Marxism as expressed through various communist writings and states. Darwinian evolution through natural selection is not intended to have political implications, but rather to describe the way that nature works. But because

8.2 Universalism

of what the Soviets viewed as imperialist and anti-communist implications, potentially dangerous to the ongoing revolution, a different view of nature was adopted as state orthodoxy in the Soviet Union with disastrous consequences. Trofim Lysenko embraced Jean-Bapstiste Lamarck's theory of evolution through the passing on not of genetic traits but of adopted characteristics. In other words, evolution could be directed by the perfection of individuals and their traits, rather than rely upon nature to weed out maladaptive, genetic characteristics in the face of a changing environment. This theory conformed better to the Hegelian notion of dialectical materialism which undergirds communist theory, and so was deemed correct, regardless of what observation actually confirms about Darwinism.

The theory was woefully mistaken, however. Lamarckian views about changes in species over time are not borne out by observation, and the genetic theory of speciation and change, and the Darwinian theory of natural selection, have continued to be confirmed by observation for more than a hundred years. But because of the closer fit between Lamarck and the Soviet view of the perfectibility of humans and the necessity for cooperation rather than competition in communist society, Lysenko rose within the Communist Party apparatus in the USSR and the genetic theory of inheritance was seen as subversive. Rather, his ideas about manipulating species by forcing changes to their phenotypes were embraced against all evidence to the contrary because of ideology. This was Soviet science, as opposed to western, decadent, capitalist science. Lysenko was supported by Stalin due to his ideology, even in the face of repeated failures in the field to successfully apply his views to crops. Thousands of geneticists who opposed Lysenko's science were jailed and even executed, and the Lenin Academy of Agricultural Sciences decreed that only Lysenkoism would be taught and not the genetic theory of inheritance. The net result was that crop yields in the USSR fell due to attempts to apply Lysenko's theories, even at a time when there were food shortages and widespread crop failures. This may well have added to the scourge of starvation that afflicted the USSR under Stalin. Moreover, Soviet science was set back by decades.

There is no "Soviet" science. The universal truths revealed by the discovery of genetic inheritance and continued confirmation of the Darwinian theory of natural selection as the mechanism for change and speciation do not depend upon local conditions, much less political opinions. The notion that a scientific theory should be prioritized due to its closer relation to a favored political ideology flies directly in the face of the universalism of science. Adopting Lysenkoism harmed society, and set back Soviet science. Soviet scientists who bravely opposed it were doing justice to the greater value of universalism in science, and literally put their lives on the line in defense of sound scientific principles. Universalism means we do not get to pick and choose among vying scientific hypotheses or theories based upon either democratic means (majority opinions) or dominant political ideologies. Rather, all researchers working in a field, regardless of where they are, their viewpoints, preferences or beliefs must accept the evidence as it comes to them and follow it where it leads. The laws of nature are the same everywhere.

This means that even where some scientific observation leads to a hypothesis that does not suit our current views, either about science or society, we are bound to test the hypothesis elsewhere, to seek either confirmation or falsification, and if confirmation continues, to revise our views. Individual, local, national, or ideological positions cannot influence our quest for the truth. When a truth becomes too "inconvenient" for a particular ideological viewpoint, or even a previously believed scientific hypothesis or theory that becomes falsified in the face of new evidence, we must adopt it nonetheless and change our viewpoint.

Society will be harmed otherwise. Whether by some physical harm, such as starvation or climate change, failure to alter our behaviors due to local beliefs or ideologies even in the face of conflicting evidence, harms society. Firstly, it places ideology above fact, above observation. Society entrusts the search for nature's laws to scientists and society has a right to know the fruits of that search regardless of politics or belief.

8.3 Communalism

Science cannot be conducted successfully in a vacuum. It depends for its proper progress upon the work of research groups throughout the world, competing in many ways, but always working toward the same goal: the truth. In order to reach that goal, and although researchers may well compete with one another for victory in the race, every stakeholder must recognize that the institutions of science are ultimately communal. In other words, every researcher works somehow in conjunction with all others, even when we compete. How is this possible?

To be scientific, a hypothesis must be testable. It must be capable of either confirmation or falsification by experiment. The experiments that confirm or falsify it must be capable of replication by others. It is vital that researchers in disparate environments, with differing backgrounds, and other than those who devised a particular hypothesis, test it. Any number of biases and errors can cause false positives or negatives, and because we know that the laws of nature are universal and not confined to any particular lab or environment, we can only eventually become more certain of a hypothesis and perhaps adopt it as a theory if others confirm it, adjusting and correcting for potential biases and errors.

For this to work, results must be published. Not only results, but methods and mechanisms too must be disclosed. In order for the communal nature of science to function properly, others must be able to replicate and reproduce all aspects of a study, and then to alter and adjust until confirmation or falsification is achieved, and then those results must become public. Hiding, obfuscating, altering, or otherwise failing to make results and methods available for other researchers to test results in pathological science – pseudoscience. If someone hides their data, fails to disclose their methods, or otherwise hinders testing elsewhere, they are defeating an essential part of its nature, they fail to act communally.

Arguably, certain market forces may be responsible for undermining the communalism of science, including the publishing of data behind paywalls. Modern researchers do have a choice to publish wherever they can in open access journals in order to increase the chances of diverse research groups to test and replicate (or falsify) their studies. Sufficient opportunities now exist to increase the communal availability of research in the broader community and to fulfill our duties to the norm of communalism.

Communalism requires attending other labs, conferences, as well as publishing, keeping up to date on the state of the art and employing best lab practices. Failing to abide by the communal nature of science harms society as well. Society is the engine that powers science, through funding and education and by vesting in scientists the trust necessary and tools available to search for truths. Not sharing the results of this work, largely done through the good graces and wealth of society, is a form of theft. Science owes society the results of its study, and this means that researchers must share with each other and society at large as much as possible of the fruits of their studies. Secreting away what is discovered in a lab not only does no service to the pursuit of the truth, it takes from the public that which is theirs.

8.4 Disinterestedness

In order to work properly, scientists must attempt to inhabit a position of equipoise, or disinterestedness. This disinterest does not mean apathy. Rather, it means that regardless of where our observations take us, we will follow them, even if they contradict our expectations. Typically, our expectations are that a hypothesis can be confirmed through some experiment, yet in most cases this is not true. Rather, hypotheses often lead us down false paths, and end up becoming falsified by some experiment, or at the very least never confirmed. But the expectation of a certain result threatens the position of equipoise. Too often, expectations and a failure therefore to maintain proper equipoise, lead to errors or worse – fraud.

Disinterestedness demands setting aside expectations, conducting experiments to test a hypothesis, and living with the results. Failing to do this dis-serves society, and may result in failures of science that can be costly or tragic. Dead ends in research should be noted and set aside so that new leads are followed, so that the truth can continue to be pursued. A failure of equipoise can lead researchers to continue to pursue dead-ends when they ought not. This wastes resources at the very least, and can deceive society at worst. Ultimately, failures like this do harm to the institutions of science and cause mistrust on the part of a wary public. This can happen even if the science is properly pursued and the hypotheses confirmed. This can happen because the public, which depends upon scientists to deliver a dispassionate account of their studies, may lose trust in scientists who appear to be motivated by some vested interest in a particular result. Even the appearance of a lack of disinterestedness should be avoided, and researchers ought to attempt to adopt an actual stance of disinterest in order to avoid the sort of bias related errors that have caused significant loss of public confidence in the past. This means being actually detached from achieving a particular result, even though we might stand something to gain (a publication, notoriety, promotion, etc.) by achieving that result. Our concern should remain with the truth.

One example that illustrates the dangers of not only being but appearing too vested in a result involves the recent "climategate" non-scandal. The gist of this involved several emails that were revealed following hacking by, apparently, a news organization. The hackers released correspondence among an international group of climate researchers. Among the thousands of emails going back and forth to keep research teams in touch and coordinating their multi-center research project, were a few emails that concerned an apparent plateau in global warming. The emails seemed to indicate that they were concerned about the way that data which seemed to indicate a plateau should best be represented so that its release would not cause the general public to doubt the still clear warming trend in the global climate. Their discussions focused on the use of various data visualizations and representations that might best portray the data in such a way that it would be clear that warming was still occurring. When the hacked emails were released, this discussion was used as the basis for a string of news stories that leveraged it to try to show that anthropogenic climate change was not occurring and that the planet was not warming. A public that was not used to internal communications among researchers geared at ensuring that data was represented in ways that the scientists preferred, were easily confused and could believe the news organizations involved when they implied that scientists were twisting the data.

No scientist sought to manipulate data, and the concerns raised among the scientists involved was not fraudulent nor did it evidence any desire to defraud. Rather, the communications were properly geared toward representing the truth to the public. The news organization that appears to have obtained the emails by hacking picked a few phrases that could be taken ambiguously if out of context, including the word "trick" in regards to a statistical method. These words may have been carelessly chosen, and the clear concern of the scientists to represent the science in such a way, truthfully, so that the public is not put in doubt about the Anthropogenic Global Warming (AGW) hypothesis. Did that conversation display a proper degree of disinterest? Arguably not. Albeit the conversation was being undertaken by private emails rather than out in public, the emails exchanged involved work accounts, using servers that were generally public property as belonging to public research institutions. As such, the scientists should not have acted as though their conversations were private. For truly private conversations, they ought to have used different media. Moreover, it is clear that the hacking of the email servers was unethical and possibly illegal. Notwithstanding, the duty of the scientists to display as well and maintain an internal position of equipoise was violated and to the detriment of climate science. The primary duty that all scientists have to the truth means that the public's sentiments and concerns, their ability or inability to understand, come second to describing nature and her truths accurately and disinterestedly. Rather than worrying and discussing the public's perceptions, put all the data out there, use the best representations, describe it and what it means, and save conversations about what amount to political concerns for private, and not correspondence on public servers.

The scientists involved have suffered more public scrutiny than they should have, and their research has been found to be sound. AGW is widely accepted and the data has been subjected to testing and confirmation. Meanwhile, public perception and trust have faltered, perhaps due to unjust reasons, but understandably given the apparent lack of equipoise displayed in the communications of the scientists involved who have nonetheless been thoroughly cleared of any wrongdoing. Science suffered, however, and a lack of equipoise, and failure to properly display disinterest or stand in the proper position of disinterest, is counter to the Mertonian norms. We have a duty to be and display disinterest. Society now will suffer for this failure as we continue to debate over an issue that scientists overwhelmingly are not confused or in disagreement about. Society has been dis-served, and we will now all potentially suffer. Our duty to science and to society requiring that the truth be discovered and properly revealed risks being obscured by our concerns with particular points of view or our interests, no matter how warranted, in a particular outcome.

8.5 Organized Skepticism

The only way that science works is if not only do individual researchers maintain a state of equipoise and understand that the current, accepted truths of science must be treated as contingent, but the whole community of researchers, the institutions of science, and the public in general understand this to be so. Skepticism means doubt. It does not mean denial. When we are confronted with a claim for which we have no evidence, we should begin from the standpoint of doubt. Only in the face of testing and the accumulation of evidence should we begin to set aside that doubt, but the testing should be systematic and conform to standard empirical methods. Moreover, even as we gain more confidence in the possible truth of a claim, even as the evidence accumulates, we remain open to the possibility that a further test may prove it to be wrong. A single falsification can be enough to topple the strongest theory, and science only functions non-pathologically when we accept this. The opposite stance is scientism, which is as counterproductive to the steady progress of science as blind faith.

Scientists owe a duty to society to continue to point out the role of doubt in science, to ensure a well-informed public understands the nature of scientific law as contingent, susceptible to revision, and built upon the methods of organized skepticism. Often it may appear to an ill-informed public that scientists are at odds with one another, that there are conflicts among them as to what constitutes the accepted "truths" of science, and even where this is not the case media reports about disagreements can make disagreements within the scientific community appear magnified. Scientists owe a duty to each other to maintain their stance of doubt, remain skeptical even about the most well established theories, be aware that the current state of scientific knowledge is always contingent, and continue to test and verify, falsify and revise the body of scientific knowledge. Failure to this, to work within institutions that adopt the position of organized skepticism both within and among other research groups, will undermine the progress of science which is a disservice to society.

As we have seen in many examples above, numerous sources of bias may unduly affect scientific judgments, may prevent properly acquiring, analyzing, or disseminating the best knowledge gleaned from the best experiments, and may arise either consciously or unconsciously. It is incumbent on every researcher to learn about the nature and sources of such biases and to do whatever they can to correct for them. This may be by education, by institutional controls, by proper experimental design, or by any number of other means. Only through awareness, though, of the fragility of scientific objectivity can we begin to correct for it properly and effectively. It is altogether too easy to believe that science, because it is ultimately self-correcting, does not require our careful, ethical concern on an individual level. But as we have also seen, individuals behaving badly can have significant deleterious effects, far-reaching and societal, and not merely individual.

From the standpoint of organized skepticism, it is proper to look for and point out not just scientific errors, but failures of the sort we have discussed in this book. Failing to abide by the Mertonian norms is arguably a failure to properly contextualize science and its methods. In numerous cases we have seen that such failures may arise to ethical lapses, or at least help cause such lapses. We must trust but verify in order for science and its institutions to function properly, and to serve society best, as we must. The alternative has always been more oversight, more regulation, and more control. And while in some cases this might be necessary, it is not preferable to scientists behaving properly in the first place. The good in science can be attained best by scientists motivated to do so, aware of the norms of properly functioning science and institutions, and eager to commit to the search for the truth using the norms of science as their guide.

Scientists who are familiar with both the history of science and the history of its faults and errors will be more able to confront their own activities, prepared better to act in ways that better conform to its norms, and less likely to need the help of other "experts" in scientific ethics and integrity should the need for ethical consideration arise.

8.6 Some Conclusions and Some Remaining Questions

I have only touched on some of the many issues that might be considered "ethical" issues in science, and we are constantly being made aware of similar concerns thanks to rapid news cycles and the ever expanding reach of science into our everyday lives. Too often, we are made aware of ethical concerns in the pursuit of science and academic research through some public failure. When that happens, the public is often, and rightly, concerned about scientists and their failures, and may unfortunately lose faith in the nature of the institutions of science. We have a duty to prevent this, not only because of our self-interest as members of its institutions, but

8.6 Some Conclusions and Some Remaining Questions

because no other institution has done so much to improve the lives of so many so rapidly and dramatically.

As I have admitted, I have addressed the issues herein not from the standpoint of any particular moral theory, although I believe that understanding something about moral theory is helpful, but rather from the standpoint from which the norms of science and its institutions that make it function as it does, imply what we have come to associate with ethical behaviors. Nothing about my approach is meant to undermine the notion that philosophical ethics is not worthwhile. On the contrary, I have given it a brief introduction in this book and advise its study. Philosophers and ethicists play an important role in helping to devise codes of behavior, ethical analysis of science, its history, and its methods, and will always do so. But I argue that the necessity of certain behaviors in science comes before ethics. Many of the right modes of acting, which we might call conforming to scientific integrity or research ethics, are necessary just for science to function. However, these behaviors might well be insufficient to be good while being a scientist. There may well be other behaviors that in themselves would not undermine science, but that may well harm others or otherwise fall outside of "the good" for any number of reasons.

The future of science is also uncertain. Its public embrace and its support from the public sphere is on shaky ground. This is for numerous political and economic reasons and not primarily due to ethical lapses by scientists. But every public ethical lapse, every case of individuals, groups, or society harmed by failures by scientists, threatens its ongoing support by the public. Moreover, any time someone is harmed, some will look for those responsible with an eye toward blame and vengeance, as well as for justice. In order to help ensure a safe future for science, we should be open to the addition of further norms. As with the contingency of scientific laws, so too are the norms of science contingent. Any one of the norms I have discussed and taken for granted here is open to testing, confirmation, or falsification. Further norms too could well be called for. The project of bringing integrity to scientific study and ethics to research remains a dynamic one. It should be discussed, debated, the principles and methods I have urged in this book are open too to debate, revision, critique, or abandonment.

The scientific community is rapidly changing and the nature of science too is dynamic, perhaps more so than ever. Because of the rapid pace of scientific and technological advance, it is incumbent upon us to stay abreast of its advances, to consider our notions about "the good" in scientific conduct and research, and engage with researchers in as many different fields as possible to comb for cases, test, and maybe revise our notions, and develop better methods of helping to create an atmosphere of scientific integrity. There is yet no good empirical evidence about how, specifically, to do this. It is a major lacuna in science that, as yet, we lack a proven method of inculcating ethical behaviors in researchers. It is evident too from the apparent yearly rise in instances of scientific and research misconduct, that something must be done. It is unfortunate that as of yet, we do not know what, specifically, we should do. This is why we must follow up with education, do some thorough research, and apply scientific means to research and teaching in the area of scientific integrity.

It will not be sufficient to have courses, handbooks, posters, or other materials that attempt to teach scientists what is "good" and ethical in scientific conduct. Only once we know that doing so, or doing so in a particular matter *actually changes behaviors*, will we have made progress in this nascent field. Which is why we would be wise to approach the problem pluralistically. Any number of approaches may work, we must try them, test them, gather data and analyse it to know if any of this sort of pedagogy and concern actually alters and improves scientific integrity. This is but one approach among many, and as good scientists, we must be open to other approaches until we start to gather some confirming evidence about them.

I have addressed duties to society in this chapter, but to close out this discussion we should consider society's duties to science. For it is through science that our modern lifestyles are possible. Never in the history of humankind has so much been available to so many, and our lifespans and levels of comfort have improved dramatically since its advent. This is true even despite intolerable levels of inequality of access to the fruits of science and technology. Society owes science, its practitioners, and its institutions plenty. We are in a state of mutual interdependence and society as it is would not exist without scientists pursuing the truths of nature. Thus, even when there are lapses, and even where science becomes set back by such lapses, it progresses in general and over time. Society owes a debt to those who work within its norms, according to ethical principles, and always with an eye toward the steady accumulation of better knowledge about the universe and its laws. By and large, the majority of scientists act ethically, whether consciously or not, and we are all thankful for their commitment and contributions to an ever progressing society, benefitting as we do both materially and intellectually from their tireless pursuits.

Study and Discussion Questions

1. Who serves whom: do scientists serve society, or does society serve scientists? What depends upon the nature of this relationship?
2. How can we best protect against the influence of political or other ideologies upon the work of science and scientists? What role do scientists play in preventing it, and what role does law and regulation play?
3. How can scientists best serve society and must they do so through direct means, or is the general accumulation of knowledge over time sufficient?
4. What role does competition have in modern science, and how can it be harnessed for the public good?

References

Birnholtz, Jeremy P., and Matthew J. Bietz. 2003. Data at work: Supporting sharing in science and engineering. In *Proceedings of the 2003 international ACM SIGGROUP conference on Supporting group work*. ACM.

Borgman, Christine L. 2012. The conundrum of sharing research data. *Journal of the American Society for Information Science and Technology* 63(6): 1059–1078.

References

Fienberg, Stephen E., Margaret E. Martin, and Miron L. Straf (eds.). 1985. *Sharing research data*. Washington, DC: National Academies.

Gordin, Michael D. 2012. How lysenkoism became pseudoscience: dobzhansky to velikovsky. *Journal of the History of Biology* 45(3): 443–468.

Grundmann, Reiner. 2013. "Climategate" and The Scientific Ethos. *Science, Technology & Human Values* 38(1): 67–93.

Holliman, Richard. 2011. Advocacy in the tail: Exploring the implications of 'climategate' for science journalism and public debate in the digital age. *Journalism* 12(7): 832–846.

Krementsov, Nikolai. 2000. Lysenkoism in Europe: Export-import of the Soviet model. *Academia in upheaval: Origins, transfers, and transformations of the Communist Academic Regime in Russia and East Central Europe*, 179–202. New York: Garland Publishing Group.

Leiserowitz, Anthony A., et al. 2013. Climategate, public opinion, and the loss of trust. *American Behavioral Scientist* 57(6): 818–837.

Lewontin, Richard, and Richard Levins. 1976. The problem of Lysenkoism. In *The radicalization of science*, ed. H. &. S. Rose. London: Macmillan.

Maibach, Edward, et al. 2012. The legacy of climategate: undermining or revitalizing climate science and policy? *Wiley Interdisciplinary Reviews: Climate Change* 3(3): 289–295.

Rai, Arti Kaur. 1999. Regulating scientific research: Intellectual property rights and the norms of science. *Northwestern University Law Review* 94: 77.

Sonneborn, T.M. 1950. Heredity, environment, and politics. *American Association for the Advancement of Science. Science* 111: 529–539.

Tenopir, Carol, et al. 2011. Data sharing by scientists: Practices and perceptions. *PloS One* 6(6): e21101.

Vickers, Andrew J. 2006. Whose data set is it anyway? Sharing raw data from randomized trials. *Trials* 7(1): 15.

Wallis, Jillian C., Elizabeth Rolando, and Christine L. Borgman. 2013. If we share data, will anyone use them? Data sharing and reuse in the long tail of science and technology. *PloS One* 8(7): e67332.

Ziman, John. 2002. *Real science: What it is and what it means*. Cambridge: Cambridge University Press.

Appendix: Codes and Principles

The Nuremberg Code

1. The voluntary consent of the human subject is absolutely essential.

   This means that the person involved should have legal capacity to give consent; should be so situated as to be able to exercise free power of choice, without the intervention of any element of force, fraud, deceit, duress, over-reaching, or other ulterior form of constraint or coercion; and should have sufficient knowledge and comprehension of the elements of the subject matter involved, as to enable him to make an understanding and enlightened decision. This latter element requires that, before the acceptance of an affirmative decision by the experimental subject, there should be made known to him the nature, duration, and purpose of the experiment; the method and means by which it is to be conducted; all inconveniences and hazards reasonably to be expected; and the effects upon his health or person, which may possibly come from his participation in the experiment.

   The duty and responsibility for ascertaining the quality of the consent rests upon each individual who initiates, directs or engages in the experiment. It is a personal duty and responsibility which may not be delegated to another with impunity.
2. The experiment should be such as to yield fruitful results for the good of society, unprocurable by other methods or means of study, and not random and unnecessary in nature.
3. The experiment should be so designed and based on the results of animal experimentation and a knowledge of the natural history of the disease or other problem under study, that the anticipated results will justify the performance of the experiment.
4. The experiment should be so conducted as to avoid all unnecessary physical and mental suffering and injury.

5. No experiment should be conducted, where there is an *apriori* reason to believe that death or disabling injury will occur; except, perhaps, in those experiments where the experimental physicians also serve as subjects.
6. The degree of risk to be taken should never exceed that determined by the humanitarian importance of the problem to be solved by the experiment.
7. Proper preparations should be made and adequate facilities provided to protect the experimental subject against even remote possibilities of injury, disability, or death.
8. The experiment should be conducted only by scientifically qualified persons. The highest degree of skill and care should be required through all stages of the experiment of those who conduct or engage in the experiment.
9. During the course of the experiment, the human subject should be at liberty to bring the experiment to an end, if he has reached the physical or mental state, where continuation of the experiment seemed to him to be impossible.
10. During the course of the experiment, the scientist in charge must be prepared to terminate the experiment at any stage, if he has probable cause to believe, in the exercise of the good faith, superior skill and careful judgement required of him, that a continuation of the experiment is likely to result in injury, disability, or death to the experimental subject.

"Trials of War Criminals before the Nuremberg Military Tribunals under Control Council Law No. 10", Vol. 2, pp. 181–182. Washington, D.C.: U.S. Government Printing Office, 1949.]

Declaration of Helsinki

Recommendations Guiding Doctors in Clinical Research

Adopted by the 18th World Medical Assembly, Helsinki, Finland, June 1964

INTRODUCTION

It is the mission of the doctor to safeguard the health of the people. His knowledge and conscience are dedicated to the fulfillment of this mission.

The Declaration of Geneva of The World Medical Association binds the doctor with the words: "The health of my patient will be my first consideration" and the International Code of Medical Ethics declares that "Any act or advice which could weaken physical or mental resistance of a human being may be used only in his interest."

Because it is essential that the results of laboratory experiments be applied to human beings to further scientific knowledge and to help suffering humanity, The World Medical Association has prepared the following recommendations as a guide to each doctor in clinical research. It must be stressed that the standards as drafted are only a guide to physicians all over the world. Doctors are not relieved from criminal, civil and ethical responsibilities under the laws of their own countries.

In the field of clinical research a fundamental distinction must be recognized between clinical research in which the aim is essentially therapeutic for a patient, and the clinical research, the essential object of which is purely scientific and without therapeutic value to the person subjected to the research.

I. BASIC PRINCIPLES

1. Clinical research must conform to the moral and scientific principles that justify medical research and should be based on laboratory and animal experiments or other scientifically established facts.
2. Clinical research should be conducted only by scientifically qualified persons and under the supervision of a qualified medical man.
3. Clinical research cannot legitimately be carried out unless the importance of the objective is in proportion to the inherent risk to the subject.
4. Every clinical research project should be preceded by careful assessment of inherent risks in comparison to foreseeable benefits to the subject or to others.
5. Special caution should be exercised by the doctor in performing clinical research in which the personality of the subject is liable to be altered by drugs or experimental procedure.

II. CLINICAL RESEARCH COMBINED WITH PROFESSIONAL CARE

1. In the treatment of the sick person, the doctor must be free to use a new therapeutic measure, if in his judgment it offers hope of saving life, reestablishing health, or alleviating suffering.

    If at all possible, consistent with patient psychology, the doctor should obtain the patient's freely given consent after the patient has been given a full explanation. In case of legal incapacity, consent should also be procured from the legal guardian; in case of physical incapacity the permission of the legal guardian replaces that of the patient.
2. The doctor can combine clinical research with professional care, the objective being the acquisition of new medical knowledge, only to the extent that clinical research is justified by its therapeutic value for the patient.

III. NON-THERAPEUTIC CLINICAL RESEARCH

1. In the purely scientific application of clinical research carried out on a human being, it is the duty of the doctor to remain the protector of the life and health of that person on whom clinical research is being carried out.
2. The nature, the purpose and the risk of clinical research must be explained to the subject by the doctor.
3a. Clinical research on a human being cannot be undertaken without his free consent after he has been informed; if he is legally incompetent, the consent of the legal guardian should be procured.
3b. The subject of clinical research should be in such a mental, physical and legal state as to be able to exercise fully his power of choice.

3c. Consent should, as a rule, be obtained in writing. However, the responsibility for clinical research always remains with the research worker; it never falls on the subject even after consent is obtained.
4a. The investigator must respect the right of each individual to safeguard his personal integrity, especially if the subject is in a dependent relationship to the investigator.
4b. At any time during the course of clinical research the subject or his guardian should be free to withdraw permission for research to be continued.
The investigator or the investigating team should discontinue the research if in his or their judgement, it may, if continued, be harmful to the individual. ∎

© **2015 *World Health Organization***

The Belmont Report

Office of the Secretary
Ethical Principles and Guidelines for the Protection of Human Subjects of Research
The National Commission for the Protection of Human Subjects of Biomedical and
  Behavioral Research
April 18, 1979

AGENCY: Department of Health, Education, and Welfare.

ACTION: Notice of Report for Public Comment.

SUMMARY: On July 12, 1974, the National Research Act (Pub. L. 93–348) was signed into law, there-by creating the National Commission for the Protection of Human Subjects of Biomedical and Behavioral Research. One of the charges to the Commission was to identify the basic ethical principles that should underlie the conduct of biomedical and behavioral research involving human subjects and to develop guidelines which should be followed to assure that such research is conducted in accordance with those principles. In carrying out the above, the Commission was directed to consider: **(i)** the boundaries between biomedical and behavioral research and the accepted and routine practice of medicine, **(ii)** the role of assessment of risk-benefit criteria in the determination of the appropriateness of research involving human subjects, **(iii)** appropriate guidelines for the selection of human subjects for participation in such research and **(iv)** the nature and definition of informed consent in various research settings.

The Belmont Report attempts to summarize the basic ethical principles identified by the Commission in the course of its deliberations. It is the outgrowth of an intensive four-day period of discussions that were held in February 1976 at the Smithsonian Institution's Belmont Conference Center supplemented by the monthly deliberations of the Commission that were held over a period of nearly four years. It is a statement of basic ethical principles and guidelines that should assist in resolving the ethical problems that surround the conduct of research with human subjects. By publishing the Report in the Federal Register, and providing reprints upon request, the Secretary intends that it may be made readily available to scientists, members of Institutional Review Boards, and Federal employees. The two-volume Appendix, containing the lengthy reports of experts and specialists who assisted the Commission in fulfilling this part of its charge, is available as DHEW Publication No. (OS) 78–0013 and No. (OS) 78–0014, for sale by the Superintendent of Documents, U.S. Government Printing Office, Washington, D.C. 20,402.

Unlike most other reports of the Commission, the Belmont Report does not make specific recommendations for administrative action by the Secretary of Health, Education, and Welfare. Rather, the Commission recommended that the Belmont Report be adopted in its entirety, as a statement of the Department's policy. The Department requests public comment on this recommendation.

National Commission for the Protection of Human Subjects of Biomedical and Behavioral Research.

Members of the Commission

Kenneth John Ryan, M.D., Chairman, Chief of Staff, Boston Hospital for Women.
Joseph V. Brady, Ph.D., Professor of Behavioral Biology, Johns Hopkins University.
Robert E. Cooke, M.D., President, Medical College of Pennsylvania.
Dorothy I. Height, President, National Council of Negro Women, Inc.
Albert R. Jonsen, Ph.D., Associate Professor of Bioethics, University of California at San Francisco.
Patricia King, J.D., Associate Professor of Law, Georgetown University Law Center.
Karen Lebacqz, Ph.D., Associate Professor of Christian Ethics, Pacific School of Religion.
*** David W. Louisell, J.D., Professor of Law, University of California at Berkeley.
Donald W. Seldin, M.D., Professor and Chairman, Department of Internal Medicine, University of Texas at Dallas.
***Eliot Stellar, Ph.D., Provost of the University and Professor of Physiological Psychology, University of Pennsylvania.
*** Robert H. Turtle, LL.B., Attorney, VomBaur, Coburn, Simmons & Turtle, Washington, D.C.
***Deceased.

Table of Contents

Ethical Principles and Guidelines for Research Involving Human Subjects
A. Boundaries Between Practice and Research
B. Basic Ethical Principles
1. Respect for Persons
2. Beneficence
3. Justice
C. Applications
1. Informed Consent
2. Assessment of Risk and Benefits
3. Selection of Subjects

Ethical Principles & Guidelines for Research Involving Human Subjects

Scientific research has produced substantial social benefits. It has also posed some troubling ethical questions. Public attention was drawn to these questions by reported abuses of human subjects in biomedical experiments, especially during the Second World War. During the Nuremberg War Crime Trials, the Nuremberg code was drafted as a set of standards for judging physicians and scientists who had conducted biomedical experiments on concentration camp prisoners. This code became the prototype of many later codes[1] intended to assure that research involving human subjects would be carried out in an ethical manner.

The codes consist of rules, some general, others specific, that guide the investigators or the reviewers of research in their work. Such rules often are inadequate to cover complex situations; at times they come into conflict, and they are frequently difficult to interpret or apply. Broader ethical principles will provide a basis on which specific rules may be formulated, criticized and interpreted.

Three principles, or general prescriptive judgments, that are relevant to research involving human subjects are identified in this statement. Other principles may also be relevant. These three are comprehensive, however, and are stated at a level of generalization that should assist scientists, subjects, reviewers and interested citizens to understand the ethical issues inherent in research involving human subjects. These principles cannot always be applied so as to resolve beyond dispute particular ethical problems. The objective is to provide an analytical framework that will guide the resolution of ethical problems arising from research involving human subjects.

This statement consists of a distinction between research and practice, a discussion of the three basic ethical principles, and remarks about the application of these principles.

Part A: Boundaries Between Practice & Research

A. *Boundaries Between Practice and Research*

It is important to distinguish between biomedical and behavioral research, on the one hand, and the practice of accepted therapy on the other, in order to know what activities ought to undergo review for the protection of human subjects of research. The distinction between research and practice is blurred partly because both often occur together (as in research designed to evaluate a therapy) and partly because notable departures from standard practice are often called "experimental" when the terms "experimental" and "research" are not carefully defined.

For the most part, the term "practice" refers to interventions that are designed solely to enhance the well-being of an individual patient or client and that have a reasonable expectation of success. The purpose of medical or behavioral practice is to provide diagnosis, preventive treatment or therapy to particular individuals.[2] By contrast, the term "research' designates an activity designed to test an hypothesis, permit conclusions to be drawn, and thereby to develop or contribute to generalizable knowledge (expressed, for example, in theories, principles, and statements of relationships). Research is usually described in a formal protocol that sets forth an objective and a set of procedures designed to reach that objective.

When a clinician departs in a significant way from standard or accepted practice, the innovation does not, in and of itself, constitute research. The fact that a procedure is "experimental," in the sense of new, untested or different, does not automatically place it in the category of research. Radically new procedures of this description should, however, be made the object of formal research at an early stage in order to determine whether they are safe and effective. Thus, it is the responsibility of medical practice committees, for example, to insist that a major innovation be incorporated into a formal research project.(3)

Research and practice may be carried on together when research is designed to evaluate the safety and efficacy of a therapy. This need not cause any confusion regarding whether or not the activity requires review; the general rule is that if there is any element of research in an activity, that activity should undergo review for the protection of human subjects.

Part B: Basic Ethical Principles

B. Basic Ethical Principles

The expression "basic ethical principles" refers to those general judgments that serve as a basic justification for the many particular ethical prescriptions and evaluations of human actions. Three basic principles, among those generally accepted in our cultural tradition, are particularly relevant to the ethics of research involving human subjects: the principles of respect of persons, beneficence and justice.

1. Respect for Persons. – Respect for persons incorporates at least two ethical convictions: first, that individuals should be treated as autonomous agents, and second, that persons with diminished autonomy are entitled to protection. The principle of respect for persons thus divides into two separate moral requirements: the requirement to acknowledge autonomy and the requirement to protect those with diminished autonomy.

An autonomous person is an individual capable of deliberation about personal goals and of acting under the direction of such deliberation. To respect autonomy is to give weight to autonomous persons' considered opinions and choices while refraining from obstructing their actions unless they are clearly detrimental to others. To show lack of respect for an autonomous agent is to repudiate that person's considered judgments, to deny an individual the freedom to act on those considered judgments, or to withhold information necessary to make a considered judgment, when there are no compelling reasons to do so.

However, not every human being is capable of self-determination. The capacity for self-determination matures during an individual's life, and some individuals lose this capacity wholly or in part because of illness, mental disability, or circumstances that severely restrict liberty. Respect for the immature and the incapacitated may require protecting them as they mature or while they are incapacitated.

Some persons are in need of extensive protection, even to the point of excluding them from activities which may harm them; other persons require little protection beyond making sure they undertake activities freely and with awareness of possible adverse consequence. The extent of protection afforded should depend upon the risk of harm and the likelihood of benefit. The judgment that any individual lacks autonomy should be periodically reevaluated and will vary in different situations.

In most cases of research involving human subjects, respect for persons demands that subjects enter into the research voluntarily and with adequate information. In some situations, however, application of the principle is not obvious. The involvement of prisoners as subjects of research provides an instructive example. On the one hand, it would seem that the principle of respect for persons requires that prisoners not be deprived of the opportunity to volunteer for research. On the other hand, under prison conditions they may be subtly coerced or unduly influenced to engage in research activities for which they would not otherwise volunteer. Respect for persons would then dictate that prisoners be protected. Whether to allow prisoners to "volunteer" or to "protect" them presents a dilemma. Respecting persons, in most hard cases, is often a matter of balancing competing claims urged by the principle of respect itself.

2. Beneficence. – Persons are treated in an ethical manner not only by respecting their decisions and protecting them from harm, but also by making efforts to secure their well-being. Such treatment falls under the principle of beneficence. The term "beneficence" is often understood to cover acts of kindness or charity that go beyond strict obligation. In this document, beneficence is understood in a stronger sense, as an obligation. Two general rules have been formulated as complementary expressions of beneficent actions in this sense: **(1)** do not harm and **(2)** maximize possible benefits and minimize possible harms.

The Hippocratic maxim "do no harm" has long been a fundamental principle of medical ethics. Claude Bernard extended it to the realm of research, saying that one should not injure one person regardless of the benefits that might come to others. However, even avoiding harm requires learning what is harmful; and, in the process of obtaining this information, persons may be exposed to risk of harm. Further, the Hippocratic Oath requires physicians to benefit their patients "according to their best judgment." Learning what will in fact benefit may require exposing persons to risk. The problem posed by these imperatives is to decide when it is justifiable to seek certain benefits despite the risks involved, and when the benefits should be foregone because of the risks.

The obligations of beneficence affect both individual investigators and society at large, because they extend both to particular research projects and to the entire enterprise of research. In the case of particular projects, investigators and members of their institutions are obliged to give forethought to the maximization of benefits and the reduction of risk that might occur from the research investigation. In the case of scientific research in general, members of the larger society are obliged to recognize the longer term benefits and risks that may result from the improvement of knowledge and from the development of novel medical, psychotherapeutic, and social procedures.

The principle of beneficence often occupies a well-defined justifying role in many areas of research involving human subjects. An example is found in research involving children. Effective ways of treating childhood diseases and fostering healthy development are benefits that serve to justify research involving children -- even when individual research subjects are not direct beneficiaries. Research also makes it possible to avoid the harm that may result from the application of previously accepted routine practices that on closer investigation turn out to be dangerous. But the role of the principle of beneficence is not always so unambiguous. A difficult ethical problem remains, for example, about research that presents more than minimal risk without immediate prospect of direct benefit to the children involved. Some have argued that such research is inadmissible, while others have pointed out that this limit would rule out much research promising great benefit to children in the future. Here again, as with all hard cases, the different claims covered by the principle of beneficence may come into conflict and force difficult choices.

3. Justice. – Who ought to receive the benefits of research and bear its burdens? This is a question of justice, in the sense of "fairness in distribution" or "what is deserved." An injustice occurs when some benefit to which a person is entitled is denied without good reason or when some burden is imposed unduly. Another way of conceiving the principle of justice is that equals ought to be treated equally. However, this statement requires explication. Who is equal and who is unequal? What considerations justify departure from equal distribution? Almost all commentators allow that distinctions based on experience, age, deprivation, competence, merit and position do sometimes constitute criteria justifying differential treatment for certain purposes. It is necessary, then, to explain in what respects people should be treated equally. There are several widely accepted formulations of just ways to distribute burdens and benefits. Each formulation mentions some relevant property on the basis of which burdens and benefits should be distributed. These formulations are **(1)** to each person an equal share, **(2)** to each person according to individual need, **(3)** to each person according to individual effort, **(4)** to each person according to societal contribution, and **(5)** to each person according to merit.

Questions of justice have long been associated with social practices such as punishment, taxation and political representation. Until recently these questions have not generally been associated with scientific research. However, they are foreshadowed even in the earliest reflections on the ethics of research involving human subjects. For example, during the 19th and early 20th centuries the burdens of serving as research subjects fell largely upon poor ward patients, while the benefits of improved medical care flowed primarily to private patients. Subsequently, the exploitation of unwilling prisoners as research subjects in Nazi concentration camps was condemned as a particularly flagrant injustice. In this country, in the 1940's, the Tuskegee syphilis study used disadvantaged, rural black men to study the untreated course of a disease that is by no means confined to that population. These subjects were deprived of demonstrably effective treatment in order not to interrupt the project, long after such treatment became generally available.

Against this historical background, it can be seen how conceptions of justice are relevant to research involving human subjects. For example, the selection of research subjects needs to be scrutinized in order to determine whether some classes (e.g., welfare patients, particular racial and ethnic minorities, or persons confined to institutions) are being systematically selected simply because of their easy availability, their compromised position, or their manipulability, rather than for reasons directly related to the problem being studied. Finally, whenever research supported by public funds leads to the development of therapeutic devices and procedures, justice demands both that these not provide advantages only to those who can afford them and that such research should not unduly involve persons from groups unlikely to be among the beneficiaries of subsequent applications of the research.

Part C: Applications

C. Applications

Applications of the general principles to the conduct of research leads to consideration of the following requirements: informed consent, risk/benefit assessment, and the selection of subjects of research.

1. Informed Consent. – Respect for persons requires that subjects, to the degree that they are capable, be given the opportunity to choose what shall or shall not happen to them. This opportunity is provided when adequate standards for informed consent are satisfied.

While the importance of informed consent is unquestioned, controversy prevails over the nature and possibility of an informed consent. Nonetheless, there is widespread agreement that the consent process can be analyzed as containing three elements: information, comprehension and voluntariness.

Information Most codes of research establish specific items for disclosure intended to assure that subjects are given sufficient information. These items generally include: the research procedure, their purposes, risks and anticipated benefits, alternative procedures (where therapy is involved), and a statement offering the subject the opportunity to ask questions and to withdraw at any time from the research. Additional items have been proposed, including how subjects are selected, the person responsible for the research, etc.

However, a simple listing of items does not answer the question of what the standard should be for judging how much and what sort of information should be provided. One standard frequently invoked in medical practice, namely the information commonly provided by practitioners in the field or in the locale, is inadequate since research takes place precisely when a common understanding does not exist. Another standard, currently popular in malpractice law, requires the practitioner to reveal the information that reasonable persons would wish to know in order to make a decision regarding their care. This, too, seems insufficient since the research subject, being in essence a volunteer, may wish to know considerably more about risks gratuitously undertaken than do patients who deliver themselves into the hand of a clinician for needed care. It may be that a standard of "the reasonable volunteer" should be proposed: the extent and nature of information should be such that persons, knowing that the procedure is neither necessary for their care nor perhaps fully understood, can decide whether they wish to participate in the furthering of knowledge. Even when some direct benefit to them is anticipated, the subjects should understand clearly the range of risk and the voluntary nature of participation.

A special problem of consent arises where informing subjects of some pertinent aspect of the research is likely to impair the validity of the research. In many cases, it is sufficient to indicate to subjects that they are being invited to participate in research of which some features will not be revealed until the research is concluded. In all cases of research involving incomplete disclosure, such research is justified only if it is clear that **(1)** incomplete disclosure is truly necessary to accomplish the goals of the research, **(2)** there are no undisclosed risks to subjects that are more than minimal, and **(3)** there is an adequate plan for debriefing subjects, when appropriate, and for dissemination of research results to them. Information about risks should never be withheld for the purpose of eliciting the cooperation of subjects, and truthful answers should always be given to direct questions about the research. Care should be taken to distinguish cases in which disclosure would destroy or invalidate the research from cases in which disclosure would simply inconvenience the investigator.

Comprehension The manner and context in which information is conveyed is as important as the information itself. For example, presenting information in a disorganized and rapid fashion, allowing too little time for consideration or curtailing opportunities for questioning, all may adversely affect a subject's ability to make an informed choice.

Because the subject's ability to understand is a function of intelligence, rationality, maturity and language, it is necessary to adapt the presentation of the information to the subject's capacities. Investigators are responsible for ascertaining that the subject has comprehended the information. While there is always an obligation to ascertain that the information about risk to subjects is complete and adequately comprehended, when the risks are more serious, that obligation increases. On occasion, it may be suitable to give some oral or written tests of comprehension.

Special provision may need to be made when comprehension is severely limited -- for example, by conditions of immaturity or mental disability. Each class of subjects that one might consider as incompetent (e.g., infants and young children, mentally disable patients, the terminally ill and the comatose) should be considered on its own terms. Even for these persons, however, respect requires giving them the opportunity to choose to the extent they are able, whether or not to participate in research. The objections of these subjects to involvement should be honored, unless the research entails providing them a therapy unavailable elsewhere. Respect for persons also requires seeking the permission of other parties in order to protect the subjects from harm. Such persons are thus respected both by acknowledging their own wishes and by the use of third parties to protect them from harm.

The third parties chosen should be those who are most likely to understand the incompetent subject's situation and to act in that person's best interest. The person authorized to act on behalf of the subject should be given an opportunity to observe the research as it proceeds in order to be able to withdraw the subject from the research, if such action appears in the subject's best interest.

Voluntariness An agreement to participate in research constitutes a valid consent only if voluntarily given. This element of informed consent requires conditions free of coercion and undue influence. Coercion occurs when an overt threat of harm is intentionally presented by one person to another in order to obtain compliance. Undue influence, by contrast, occurs through an offer of an excessive, unwarranted, inappropriate or improper reward or other overture in order to obtain compliance. Also, inducements that would ordinarily be acceptable may become undue influences if the subject is especially vulnerable.

Unjustifiable pressures usually occur when persons in positions of authority or commanding influence -- especially where possible sanctions are involved -- urge a course of action for a subject. A continuum of such influencing factors exists, however, and it is impossible to state precisely where justifiable persuasion ends and undue influence begins. But undue influence would include actions such as manipulating a person's choice through the controlling influence of a close relative and threatening to withdraw health services to which an individual would otherwise be entitle.

2. Assessment of Risks and Benefits. – The assessment of risks and benefits requires a careful arrayal of relevant data, including, in some cases, alternative ways of obtaining the benefits sought in the research. Thus, the assessment presents both an opportunity and a responsibility to gather systematic and comprehensive information about proposed research. For the investigator, it is a means to examine whether the proposed research is properly designed. For a review committee, it is a method for determining whether the risks that will be presented to subjects are justified. For prospective subjects, the assessment will assist the determination whether or not to participate.

The Nature and Scope of Risks and Benefits The requirement that research be justified on the basis of a favorable risk/benefit assessment bears a close relation to the principle of beneficence, just as the moral requirement that informed consent be obtained is derived primarily from the principle of respect for persons. The term "risk" refers to a possibility that harm may occur. However, when expressions such as "small risk" or "high risk" are used, they usually refer (often ambiguously) both to the chance (probability) of experiencing a harm and the severity (magnitude) of the envisioned harm.

The term "benefit" is used in the research context to refer to something of positive value related to health or welfare. Unlike, "risk," "benefit" is not a term that expresses probabilities. Risk is properly contrasted to probability of benefits, and benefits are properly contrasted with harms rather than risks of harm. Accordingly, so-called risk/benefit assessments are concerned with the probabilities and magnitudes of possible harm and anticipated benefits. Many kinds of possible harms and benefits need to be taken into account. There are, for example, risks of psychological harm, physical harm, legal harm, social harm and economic harm and the corresponding benefits. While the most likely types of harms to research subjects are those of psychological or physical pain or injury, other possible kinds should not be overlooked.

Risks and benefits of research may affect the individual subjects, the families of the individual subjects, and society at large (or special groups of subjects in society). Previous codes and Federal regulations have required that risks to subjects be outweighed by the sum of both the anticipated benefit to the subject, if any, and the anticipated benefit to society in the form of knowledge to be gained from the research. In balancing these different elements, the risks and benefits affecting the immediate research subject will normally carry special weight. On the other hand, interests other than those of the subject may on some occasions be sufficient by themselves to justify the risks involved in the research, so long as the subjects' rights have been protected. Beneficence thus requires that we protect against risk of harm to subjects and also that we be concerned about the loss of the substantial benefits that might be gained from research.

The Systematic Assessment of Risks and Benefits It is commonly said that benefits and risks must be "balanced" and shown to be "in a favorable ratio." The metaphorical character of these terms draws attention to the difficulty of making precise judgments. Only on rare occasions will quantitative techniques be available for the scrutiny of research protocols. However, the idea of systematic, nonarbitrary analysis of risks and benefits should be emulated insofar as possible. This ideal requires those making decisions about the justifiability of research to be thorough in the accumulation and assessment of information about all aspects of the research, and to consider alternatives systematically. This procedure renders the assessment of research more rigorous and precise, while making communication between review board members and investigators less subject to misinterpretation, misinformation and conflicting judgments. Thus, there should first be a determination of the validity of the presuppositions of the research; then the nature, probability and magnitude of risk should be distinguished with as much clarity as possible. The method of ascertaining risks should be explicit, especially where there is no alternative to the use of such vague categories as small or slight risk. It should also be determined whether an investigator's estimates of the probability of harm or benefits are reasonable, as judged by known facts or other available studies.

Finally, assessment of the justifiability of research should reflect at least the following considerations: **(i)** Brutal or inhumane treatment of human subjects is never morally justified. **(ii)** Risks should be reduced to those necessary to achieve the research objective. It should be determined whether it is in fact necessary to use human subjects at all. Risk can perhaps never be entirely eliminated, but it can often be reduced by careful attention to alternative procedures. **(iii)** When research involves significant risk of serious impairment, review committees should be extraordinarily insistent on the justification of the risk (looking usually to the likelihood of benefit to the subject -- or, in some rare cases, to the manifest voluntariness of the participation). **(iv)** When vulnerable populations are involved in research, the appropriateness of involving them should itself be demonstrated. A number of variables go into such judgments, including the nature and degree of risk, the condition of the particular population involved, and the nature and level of the anticipated benefits. **(v)** Relevant risks and benefits must be thoroughly arrayed in documents and procedures used in the informed consent process.

3. Selection of Subjects. – Just as the principle of respect for persons finds expression in the requirements for consent, and the principle of beneficence in risk/benefit assessment, the principle of justice gives rise to moral requirements that there be fair procedures and outcomes in the selection of research subjects.

Justice is relevant to the selection of subjects of research at two levels: the social and the individual. Individual justice in the selection of subjects would require that researchers exhibit fairness: thus, they should not offer potentially beneficial research only to some patients who are in their favor or select only "undesirable" persons for risky research. Social justice requires that distinction be drawn between classes of subjects that ought, and ought not, to participate in any particular kind of research, based on the ability of members of that class to bear burdens and on the appropriateness of placing further burdens on already burdened persons. Thus, it can be considered a matter of social justice that there is an order of preference in the selection of classes of subjects (e.g., adults before children) and that some classes of potential subjects (e.g., the institutionalized mentally infirm or prisoners) may be involved as research subjects, if at all, only on certain conditions.

Injustice may appear in the selection of subjects, even if individual subjects are selected fairly by investigators and treated fairly in the course of research. Thus injustice arises from social, racial, sexual and cultural biases institutionalized in society. Thus, even if individual researchers are treating their research subjects fairly, and even if IRBs are taking care to assure that subjects are selected fairly within a particular institution, unjust social patterns may nevertheless appear in the overall distribution of the burdens and benefits of research. Although individual institutions or investigators may not be able to resolve a problem that is pervasive in their social setting, they can consider distributive justice in selecting research subjects.

Some populations, especially institutionalized ones, are already burdened in many ways by their infirmities and environments. When research is proposed that involves risks and does not include a therapeutic component, other less burdened classes of persons should be called upon first to accept these risks of research, except where the research is directly related to the specific conditions of the class involved. Also, even though public funds for research may often flow in the same directions as public funds for health care, it seems unfair that populations dependent on public health care constitute a pool of preferred research subjects if more advantaged populations are likely to be the recipients of the benefits.

One special instance of injustice results from the involvement of vulnerable subjects. Certain groups, such as racial minorities, the economically disadvantaged, the very sick, and the institutionalized may continually be sought as research subjects, owing to their ready availability in settings where research is conducted. Given their dependent status and their frequently compromised capacity for free consent, they should be protected against the danger of being involved in research solely for administrative convenience, or because they are easy to manipulate as a result of their illness or socioeconomic condition.

---

(1) Since 1945, various codes for the proper and responsible conduct of human experimentation in medical research have been adopted by different organizations. The best known of these codes are the Nuremberg Code of 1947, the Helsinki Declaration of 1964 (revised in 1975), and the 1971 Guidelines (codified into Federal Regulations in 1974) issued by the U.S. Department of Health, Education, and Welfare Codes for the conduct of social and behavioral research have also been adopted, the best known being that of the American Psychological Association, published in 1973.

(2) Although practice usually involves interventions designed solely to enhance the well-being of a particular individual, interventions are sometimes applied to one individual for the enhancement of the well-being of another (e.g., blood donation, skin grafts, organ transplants) or an intervention may have the dual purpose of enhancing the well-being of a particular individual, and, at the same time, providing some benefit to others (e.g., vaccination, which protects both the person who is vaccinated and society generally). The fact that some forms of practice have elements other than immediate benefit to the individual receiving an intervention, however, should not confuse the general distinction between research and practice. Even when a procedure applied in practice may benefit some other person, it remains an intervention designed to enhance the well-being of a particular individual or groups of individuals; thus, it is practice and need not be reviewed as research.

(3) Because the problems related to social experimentation may differ substantially from those of biomedical and behavioral research, the Commission specifically declines to make any policy determination regarding such research at this time. Rather, the Commission believes that the problem ought to be addressed by one of its successor bodies.